Handbooks for the Identification of British Insects
Vol. 6. Part 6

W0071656

The Vespoid Wasps

(Tiphiidae, Mutillidae, Sapygidae, Scoliidae and Vespidae) of the British Isles

Michael E. Archer
York St Johns University
and BWARS

Whole-insect figures by

James Turner

Amgueddfa Cymru –
National Musuem of Wales

© Royal Entomological Society, 2014.

Published for the Royal Entomological Society
The Mansion House
Bonehill
Chiswell Green Lane
Chiswell Green
St Albans
AL2 3NS
www.royensoc.co.uk

By the Field Studies Council
Unit C1
Stafford Park 15
Telford
TF3 3BB
www.field-studies-council.org

ISBN: 978 0 901546 98 2

All rights reserved. No part of this book may be
reproduced or translated in any form or by any means,
electronically, mechanically, by photocopying or
otherwise, without written permission from the
copyright holders.

Contents

Abstract

A general introduction is provided to the natural history and external morphology for the following families of the Vespoidea: Tiphiidae, Mutillidae, Sapygidae, Scoliidae and Vespidae including the subfamilies Eumeninae, Polistinae and Vespinae. A checklist is given for the 45 species that have been recorded from the British Isles. Keys to the families, genera and species, together with diagnostic illustrations, are provided. Notes are given on the preparation of specimens for identification. Species profiles include information on British and overseas distribution, habitats, adult activity period, nesting characteristics, prey, host records and flowers visited.

Acknowledgements

I would particularly like to thank Dr. M. Wilson who greatly improved the text besides suggesting changes to some of the figures. Thanks also go to the many people who provided records from which the geographical distribution of species can be known. I would also like to thank the staff of the National Museum of Wales for providing the coloured images of species.

Introduction

The aculeate Hymenoptera consists of three superfamilies: Chrysidoidea, Vespoidea and Apoidea (Gauld & Bolton, 1996). The Vespoidea, worldwide, consists of twelve families (Gauld & Bolton, 1996) although Brothers & Finnamore (*in* Goulet & Huber, 1993) indicate ten families. This Handbook deals with the following families of the Vespoidea: Tiphiidae with three species, Mutillidae with three species, Sapygidae with two species, Scoliidae with one species and the Vespidae consisting of the Eumeninae with 25 species, Polistinae with two species and the Vespinae with twelve species.

Table 1. Distribution of species that have been recorded as resident species.

	England	Wales	Scotland	Ireland	Channel Islands
Mutilla europaea	•		•		
Smicromyrme rufipes	•				•
Myrmosa atra	•	•		•	•
Monosapyga clavicornis	•	•			
Sapyga quinquepunctata	•	•			
Scolia sexmaculata					•
Tiphia femorata	•	•			•
Tiphia minuta	•	•	•	•	
Methocha articulata	•	•			
Ancistrocerus antilope	•	•	•		•
Ancistrocerus gazella	•	•	•	•	•
Ancistrocerus nigricornis	•	•	•	•	•
Ancistrocerus oviventris	•	•	•	•	•
Ancistrocerus parietinus	•	•	•	•	•
Ancistrocerus parietum	•	•	•	•	•
Ancistrocerus quadratus	•				
Ancistrocerus scoticus	•	•	•	•	
Ancistrocerus trifasciatus	•	•	•	•	•
Eumenes coarctatus	•	•			
Euodynerus quadrifasciatus	•				
Gymnomerus laevipes	•	•			
Microdynerus exilis	•				
Odynerus melanocephalus	•	•			
Odynerus reniformis	•				•
Odynerus simillimus	•				
Odynerus spinipes	•	•	•	•	•
Pseudepipona herrichii	•				
Pterocheilus phaleratus					•
Symmorphus bifasciatus	•	•	•	•	•
Symmorphus connexus	•				
Symmorphus crassicornis	•	•			
Symmorphus gracilis	•	•			
Polistes dominula	•				
Dolichovespula media	•	•	•		
Dolichovespula norwegica	•	•	•	•	
Dolichovespula saxonica	•	•			
Dolichovespula sylvestris	•	•	•	•	•
Vespa crabro	•	•			•
Vespula austriaca	•	•	•	•	
Vespula germanica	•	•	•	•	•
Vespula rufa	•	•	•	•	•
Vespula vulgaris	•	•	•	•	•

Early accounts of the tiphiids, mutillids and sapygids were given by Shuckard (1837) which included a key to the genera with species descriptions, some natural history observations and sites where the species were found. Smith (1858) considered the tiphiids, mutillids, sapygids, eumenines and vespines in terms of species morphology, natural history and sites where found but no keys were provided. No reference was made to the scoliids and polistines in these two early accounts. Saunders (1896) provided keys to species with illustrations and species morphology, description, natural history and sites where found were again considered. Additional notes on collecting specimens and preparing them for identification were also given.

Yarrow (1954) provided a key, with illustrations, to separate the newly discovered species pair *Ancistrocerus gazella* and *A. parietum*. Guichard (1972) published a key, without illustrations, to the species of *Symmorphus* in relation his discovery of *S. crassicornis*.

Spradbery (1973) provided illustrated keys to the species of eumenines and vespines and gave a full account of their morphology and natural history. He also provided much information on how to find and trap specimens and the first maps of the British Isles distribution of each species to the spatial scales of 10 km squares. Archer (1979) published many more records of vespine species and produced maps, again at 10 km square resolution, but with time categories of 1900-1949 and 1950 onwards, and provided an illustrated key to vespine species. Edwards (1980) also provided an illustrated key to vespine species. Richards (1980) provided illustrated keys to all the species considered in this Handbook including *Polistes dominula*, but not *P. gallicus*, and the scoliids. He also included some whole species drawings.

A key to the British aculeate families with illustrations, including the scoliids, was given by Gauld & Bolton (1996) in their book dealing mainly with British Hymenoptera. They also gave morphological and natural history accounts of families with whole species drawings including the scoliids and making reference to the Channel Island species, *Scolia sexmaculata*. Goulet & Huber (1993) also provide illustrated keys to the families of the Hymenoptera but in a world context.

The Bees, Wasps and Ants Recording Society (BWARS) started their Provisional Atlases in 1997 and have considered nearly all the species covered in this Handbook. Information is given for each species on nomenclature, distribution overseas and within the British Isles (with a map at 10 km square detail and three date classes of before 1900, 1900-1969 and 1970 onwards), status in Britain, habitats, adult activity period, prey collected or hosts parasitized, nesting biology, flowers visited and parasites. Archer (2003a, 2nd ed.) updated information on the eumenines and provided illustrated keys to the species. Archer (2003a) also considered the vagrant species *Eumenes papillarius* and *Rhynchium oculatum*, besides *Pterocheilus phaleratus*, which is only recorded from the Channel Islands. In the BWARS Members' Handbook (Archer, 2004, ed.) an illustrated key was given to the European Vespinae.

Status and conservation

The UK Biodiversity Group (1999) Action Plans (BAPs) for invertebrates included one eumenine species, *Pseudepipona herrichii*. This species was first studied by Morrison (1991) and Edwards & Roberts (1995) in Pre-recovery Project reports to English Nature. These reports led to autecological studies and management advice and action by Roberts (1998, 1999, 2000, 2001) and an article for British Wildlife (Roberts & Else, 1997) entitled "The Purbeck Mason-wasp – back from the brink?".

The Aculeate Conservation Group (ACG) was initiated during 1999 with funding from English Nature. Organised by Project Officer, M. Edwards, the ACG was able to sponsor work on autecological studies and management proposals for several species. With funding from the Esmée Fairbairn Foundation, ACG became Hymettus Ltd with P. Lee as the new Project Officer and M. Edwards the Secretary. The studies started by the ACG were continued by Hymettus Ltd and new studies were started resulting in the following reports on *Odynerus simillimus* (Harvey, 2001, 2002; Lee & Scott, 2007; Strudwick & Scott, 2008; Scott & Strudwick, 2009; Strudwick & Lee, 2010), *O. melanocephalus* (Falk, 2005; Hunnisett, 2006; Edwards, 2007; Wright, 2007; Wright, 2010), *Euodynerus quadrifasciatus* (Else, 2006; Else & Roberts, 2006) and *Pseudepipona herrichii* (Dieck & Neal, 2007; Dieck, 2008, 2010; Roberts, 2008). These reports are available on the Hymettus website (www.hymettus.org.uk). Some of the findings from these reports have been included in the species accounts.

The Joint Nature Conservation Committee (JNCC) asked Hymettus Ltd to work with BWARS to produce a National Review of Bee, Ant and Wasp species in Great Britain with the aim of revising the list of BAP species. The new BAP species list for England is *Pseudepipona herrichii*, *Odynerus melanocephalus* and *O. simillimus* and the key species for Scotland are *Ancistrocerus parietum*, *Symmorphus bifasciatus* and *Mutilla europaea* and for Wales *Odynerus melanocephalus*.

Of the species resident in England, Wales and Scotland, two are probably now extinct (*Ancistrocerus quadratus* and *Odynerus reniformis*), six RDB species are threatened as being rare or very rare (*Ancistrocerus antilope*, *Euodynerus quadrifasciatus*, *Odynerus simillimus*, *Pseudepipona herrichii*, *Symmorphus connexus* and *S. crassicornis*) and eight species are near-threatened, being scarce (*Mutilla europaea*, *Smicromyrme rufipes*, *Monosapyga clavicornis*, *Tiphia minuta*, *Methocha articulata*, *Eumenes coarctatus*, *Microdynerus exilis*, *Odynerus melanocephalus*). *Tiphia minuta*, which is difficult to find, is probably undergoing an increase in range, so may soon have to be removed from the near-threatened list. In contrast, *Gymnomerus laevipes* is probably undergoing a decrease in range and may be added to the near-threatened list. In conclusion, over one-third of resident species are extinct, threatened or near-threatened. In addition, *Ancistrocerus nigricornis* and *A. oviventris* have undergone a significant decrease of range in recent times while the newly introduced *Dolichovespula media* and *D. saxonica* are showing an increase in range.

Economic importance

The economic importance of the species considered here relate mainly to the Vespinae. Edwards (1980) dealt in detail with the beneficial and harmful effects of the social wasps and gave information on their control both in the home and industry. The stings of social wasps can cause pain and in a very few cases even death. Death is usually due to an allergic reaction to the venom injected by the sting. In England and Wales between 1949 and 1969, on average, three to four persons died per year from wasp stings (Edwards, 1980). Spradbery (1973) gives more detail on allergic reactions and Akre *et al.* (1980) give information on treatment to desensitise a person from the venom of the sting.

Social wasps have also been found to carry bacteria of *Escherichia coli* and *Salmonella*, which can cause gastro-enteritis and food poisoning (Edwards, 1980) and should not be allowed to walk over food (as can be observed in butcher's and cake shops). The bacteria can also enter the body during stinging.

Economic losses can be experienced in the field from fruit crop losses, particularly of plums, pears and grapes. In the factory they may be a considerable problem in icing rooms of bakeries and fruit canning factories.

Beneficial gains, particularly of mason and social wasps, are due to their predatory habits in collecting food for their larvae and in the pollination of flowers while collecting nectar. The social wasps often visit flowers such as figwort, cotoneaster, bilberry and crowberry and such flowers have been called "wasp flowers" (Proctor & Yeo, 1973; Proctor, Yeo & Lack, 1996). Social wasps kill large numbers of insects and spiders, particularly adult flies and lepidopteran larvae (Archer, 1977). Archer (2005) estimated that the workers of a single large colony of *Vespula vulgaris* made in excess of half a million foraging trip for prey during its colony's existence probably resulting in an unknown saving in the use of insecticides and an increase in crop production.

External morphology

The body of an insect which is covered by a number of plates, or **sclerites,** is generally divided into three parts called the **head, thorax** and **abdomen**. The thorax and abdomen are usually clearly made up of a number of segments. In the Hymenoptera, the first abdominal segment has been become associated with the thorax so that the names, thorax and abdomen, are no longer used. The middle part of the body can now be called the **mesosoma** and the hind part the **gaster** or **metasoma**. The following morphological description is most readily shown by the Vespidae. For the other families, particularly for wingless females, the external morphology can be modified. Some of these modifications are given in the keys.

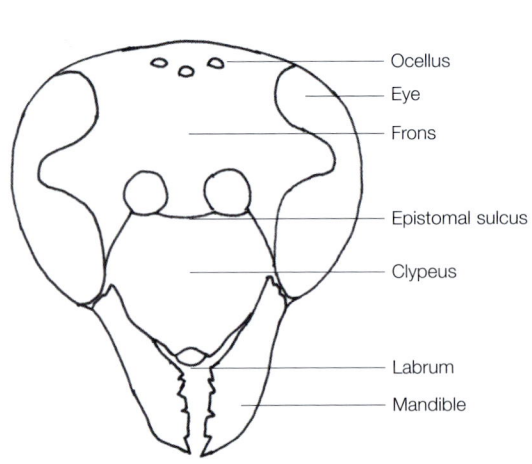

Figure 1. Frontal view of head.

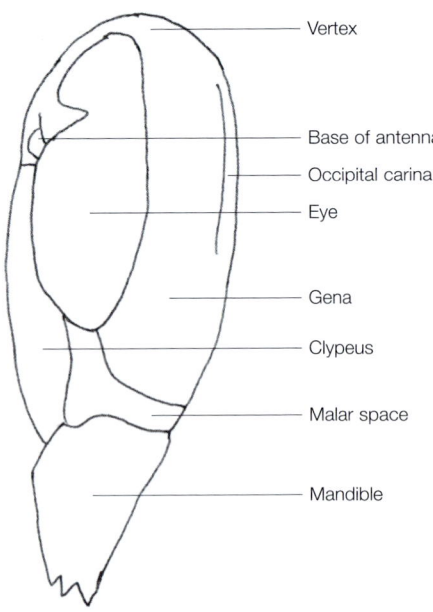

Figure 2. Lateral view of head.

The plates or sclerites that cover the head (Figs 1, 2) are fused together to form a capsule although it is joined to the mesosoma by a flexible neck. The top of the head is called the **vertex** on which are positioned three **ocelli** arranged in a triangle with two lateral ocelli and a median ocellus. The sides of the head are occupied by the **eyes**. The area between the eyes is called the **frons** and the area behind the compound eyes forming the sides of the head the **genae**. Below the frons is the **clypeus,** below which the **labrum** may be visible. The dorsal margin of the clypeus may be marked by a depression called the **epistomal sulcus**. Attached to the underside of the clypeus is the **labium** which may be concealed.

Ventrally are the **mouthparts** with a pair of **mandibles**. The space between the mandible and eye is called the **malar space**. Further structural details of the mouthparts and the back of the head around the **foramen magnum** where the head joins the mesosoma via the flexible neck are not considered as they are not required for identification purposes. In addition, ridges or **carinae** are also present, of which the **occipital carina,** which defines the posterior margin of the gena, should be noted. Further structural details can be found in Gauld & Bolton (1996).

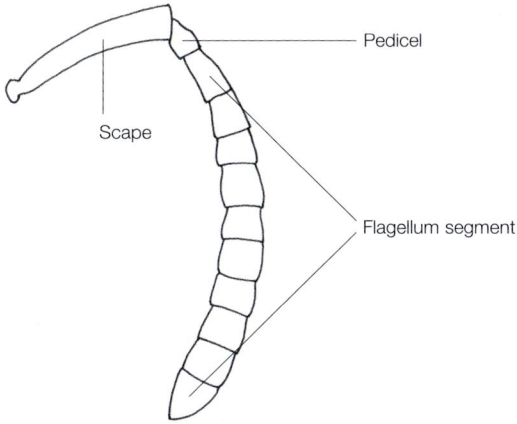

Figure 3. View of antenna.

The **antennae** (Fig. 3) arise from a circular depression, or **antennal sulcus**. The antennae which are mobile in their sockets consist of a basal **scape**, followed by a usually small **pedicel** segment and the **flagellum** of ten or more segments. Sometimes the pedicel and flagellum segments are similar to each other. The undersides of segments of the flagellum may bear one or two swellings or **tyloid(s).**

The sclerites of the mesosoma (Figs 4, 5) are generally fused together to form a capsule. The mesosoma can be divided into the anterior **prothorax**, followed by the **mesothorax**, **metathorax** and **propodeum**, which is really the first segment of the abdomen. Pairs of **legs** are attached to the prothorax, mesothorax and metathorax and a pair of membranous **wings** to the mesothorax and metathorax. The prothorax consists of the **pronotum** and **propleuron**. The pronotum extends ventrally downwards and dorsally backwards to the base of the wings which are covered by the **tegulae**. Near the tegulae the pronotum may form a swollen **pronotal tubercle**. The propleuron occupies an anterior-ventral position and is loosely attached to the pronotum.

The mesothorax is divided into the dorsal mesonotum, composed of the anterior **mesoscutum** and posterior **mesoscutellum**, and the ventro-lateral **mesopleuron**. The mesopleuron may be divided into the larger anterior-ventral **mesepisternum** and the smaller posterio-dorsal **mesepimeron** which may be absent. The mesepisternum may be further divided into dorsal and ventral parts.

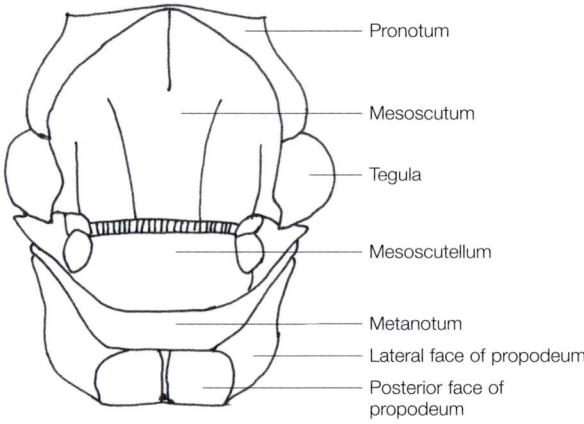

Figure 4. Dorsal view of mesosoma.

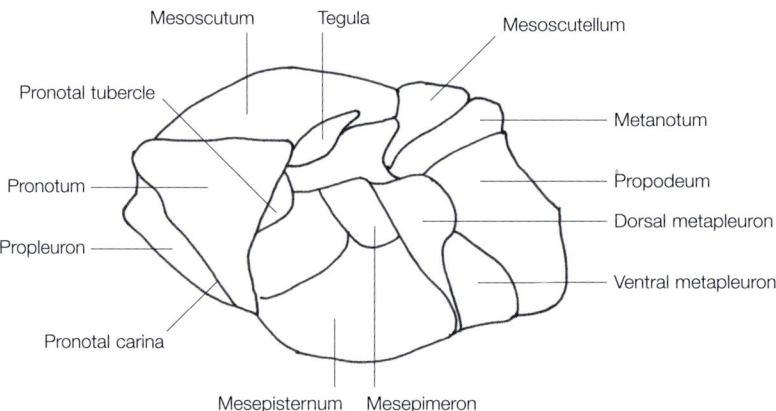

Figure 5. Lateral view of mesosoma.

The metathorax is divided into the dorsal **metanotum** and the ventro-lateral **metapleuron** which may be further divided into dorsal and ventral parts.

The propodeum consists of a single plate or sclerite. In the Eumeninae the posterior face is often separated from the rest of the propodeum by ridges. In the genus *Sapyga* the dorsal and posterior faces of the propodeum are almost at right angles to each other and separated by a ridge. The gaster articulates with the propodeum via flexible socket.

Each leg (Fig. 6) consists of a **coxa** which articulates with the thorax, followed by the **trochanter**, the **femur**, the **tibia** and the five segmented **tarsus**. The end segment of the tarsus ends in two **claws** with a soft pad or **arolium** between the claws.

The fore wing and the smaller hind wing can be held together by a row of hooks or **hamuli** along the front edge of the hind wing engaging a narrow **frenal fold** on the hind margin of the fore wing. The wings are supported by **veins** which divide the wing into closed and open

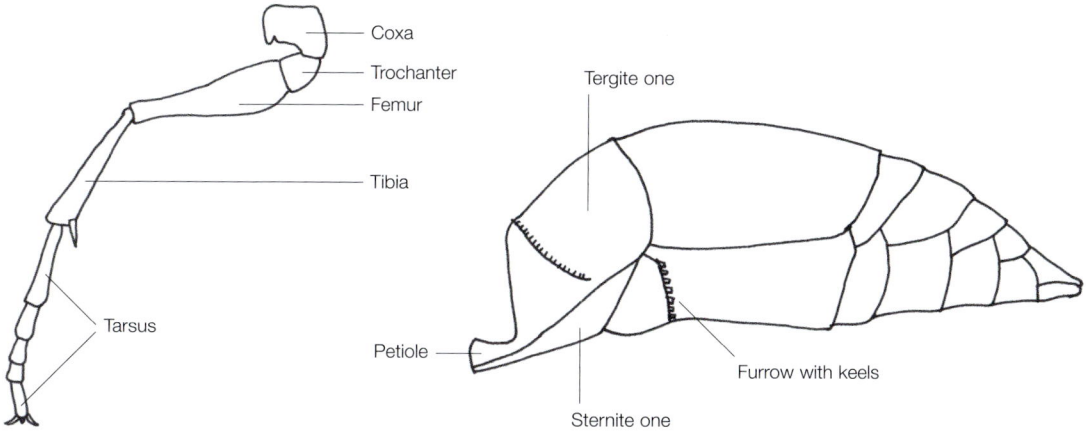

Figure 6. View of the leg.

Figure 7. Lateral view of the male gaster.

cells. Closed cells are completely surrounded by veins while open cells are not entirely surrounded by veins. The names of the veins and cells are given in the keys as required. The nomenclature of the veins and cells are those given in Gauld & Bolton (1996).

The gaster (Fig. 7) consists of a number of segments, each consisting of a dorsal **tergal** and a ventral **sternal** plate. The first gastral segment is joined to the propodeum by a **petiole**. The first tergite and sternite are attached to each other while the other tergites and sternites are movable against each other. Except for the Sapygidae, the first and second sternites are separated by a hollow or constriction. In most Eumeninae the second sternite has a basal furrow or sulcature.

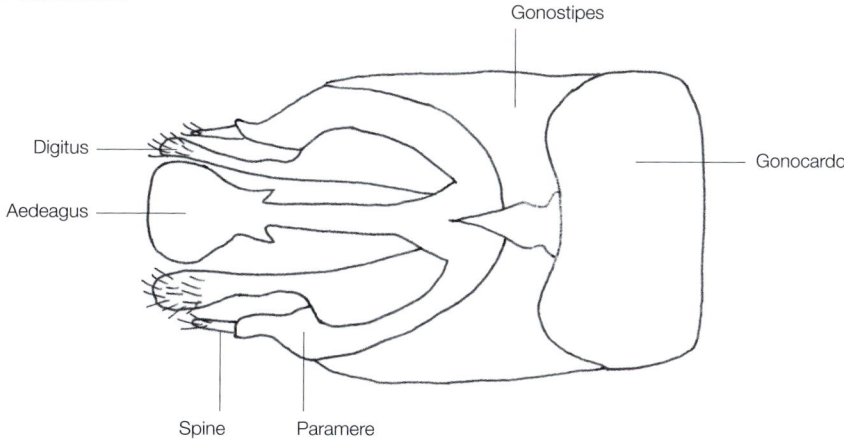

Figure 8. Ventral view of male genitalia of *Vespula vulgaris*.

The structure of the female genitalia is not used in the keys so will not be considered further. The structure of the male **genitalia** (Fig. 8) is used in the key to the Vespinae so will be considered. The male genitalia consist of a basal ring plate or **gonocardo** which encloses a pair of plates or **gonostipes** which extends posteriorly as the **parameres** which bear a **spine**. From the postero-ventral surface of each gonostipes are lobes called the **volsellae** which extend posteriorly, being free and divided into a **digitus** and **cuspis**. The volsella and cuspis are normally hidden from view and dissection is needed to see them. Inside the gonostipes is the elongate penis or **aedeagus**.

Checklist

The following checklist consists of resident species and non-resident species or vagrants. The vagrant species are *Eumenes papillarius, Rhynchium oculatum, Polistes gallicus* and possibly *Pterocheilus phaleratus*. *Polistes dominula* might be regarded as a vagrant species. However, since it has survived at the same site for a number of years with successful overwintering and showed an increase in the number of colonies, it may now be becoming a resident species. *P. dominula* has also appeared at several geographical sites sometimes showing nesting activity. *Scolia sexmaculata* and *Pterocheilus phaleratus* have only been recorded from the Channel Islands. *Vespa velutina* is considered as a potential invasion species from France.

Family MUTILLIDAE
    Subfamily Mutillinae
        MUTILLA Linnaeus, 1758
            europaea Linnaeus, 1758
        SMICROMYRME Thomson, 1870
            rufipes (Fabricius, 1787)
    Subfamily Myrmosinae
        MYRMOSA Latreille, 1796
            atra Panzer, 1801
                melanocephala (Fabricius, 1793)

Family SAPYGIDAE
        MONOSAPYGA Pic, 1920
            clavicornis (Linnaeus, 1758)
        SAPYGA Latreille, 1796
            quinquepunctata (Fabricius, 1781)

Family SCOLIIDAE
        SCOLIA Fabricius, 1775
            sexmaculata Müller, 1766 (Channel Islands)
                quadripunctata Fabricius, 1775

Family TIPHIIDAE
    Subfamily Tiphiinae
        TIPHIA Fabricius, 1775
            femorata Fabricius, 1775
            minuta Van der Linden, 1827
    Subfamily Methochinae
        METHOCHA Latreille, 1804
            articulata Latreille, 1792
                 ichneumonides Latreille, 1804

Family VESPIDAE
    Subfamily EUMENINAE
        ANCISTROCERUS Wesmael, 1836
            antilope (Panzer, 1798)
            gazella (Panzer, 1798)
            nigricornis (Curtis, 1826)
                 callosus (Thomson, 1870)

oviventris (Wesmael, 1836)
    pictus (Curtis, 1829)
parietinus (Linnaeus, 1758)
parietum (Linnaeus, 1758)
quadratus (Panzer, 1799)
    claripennis (Thomson, 1874)
scoticus (Curtis, 1826)
    trimarginatus misidentified
    albotricinctus (Zetterstedt, 1838)
trifasciatus (Müller, 1776)
    trimarginatus (Zetterstedt, 1838)

EUMENES Latreille, 1802
coarctatus (Linnaeus, 1758)
papillarius (Christ, 1791) (Vagrant)

EUODYNERUS Dalla Torre, 1904
quadrifasciatus (Fabricius, 1793)
    tomentosus (Thomson, 1870)

GYMNOMERUS Blüthgen, 1938
laevipes (Shuckard, 1837)

MICRODYNERUS Thomson, 1874
exilis (Herrich-Schäffer, 1839)

ODYNERUS Latreille, 1802
melanocephalus (Gmelin *in* Linnaeus, 1790)
reniformis (Gmelin *in* Linnaeus, 1790)
simillimus Morawitz, 1867
spinipes (Linnaeus, 1758)

PSEUDEPIPONA de Saussure, 1856
herrichii (de Saussure, 1855)
    basalis (Smith, 1857)
    variegata misidentified

PTEROCHEILUS Klug, 1805
phaleratus (Panzer, 1797) (Channel Islands)

RHYNCHIUM Spinola, 1806
oculatum (Fabricius, 1781) (Vagrant)

SYMMORPHUS Wesmael, 1836
bifasciatus (Linnaeus, 1761)
    mutinensis (Baldeni, 1894)
    sinuatissimus Richards, 1935
    sinuatus (Fabricius, 1793)
connexus (Curtis, 1826)
    bifasciatus misidentified
crassicornis (Panzer, 1798)
gracilis (Brullé, 1832)
    elegans (Wesmael, 1833)

Subfamily POLISTINAE
POLISTES Latreille, 1802
dominula (Christ, 1791)
    gallicus misidentified
gallicus (Linnaeus, 1767) (Vagrant)
    foederata Kohl, 1898
    omissa Weyrauch, 1938

Subfamily VESPINAE
   DOLICHOVESPULA Rohwer, 1916
      PSEUDOVESPULA Bischoff, 1931
      BOREOVESPULA Blüthgen, 1943
      METAVESPULA Blüthgen, 1943
      media (Retzius, 1783)
      norwegica (Fabricius, 1781)
         britannica Leach, 1814
         norwegica Oliver, 1792, misspelling
      saxonica (Fabricius, 1793)
      sylvestris (Scopoli, 1763)
   VESPA Linnaeus, 1758
      crabro Linnaeus, 1758
      velutina Lepeletier, 1836 (Potential invasion species)
   VESPULA Thomson, 1869
      PSEUDOVESPA Schmiedeknecht, 1881
      PARAVESPULA Blüthgen, 1938
      ALLOVESPULA Blüthgen, 1943
      austriaca (Panzer, 1799)
         borealis Smith, 1843
      germanica (Fabricius, 1793)
      rufa (Linnaeus, 1758)
      vulgaris (Linnaeus, 1758)

It is convenient to consider *Vespula austriaca* and *V. rufa* as the *V. rufa* species group (*Vespula* s.s.) and *V. germanica* and *V. vulgaris* as the *V. vulgaris* species group (*Paravespula*).

Lifecycle terminology

Cleptoparasite. The larval stage of a parasite that eats the food store of its host.

Diapause. Delayed development usually during the winter.

Parasite. A term used when it is unknown whether the species is a parasitoid or a cleptoparasite.

Parasitoid. The larval stage of a parasite eats the whole, or part, of its host.

Mutivoltine. More than one generation per year.

Synchronised life cycle. In temperate climates life cycle of a species start at the same time of year.

Univoltine. One generation per year.

Unsynchronised life cycle. In tropical climates life cycle of a species start at different times of the year.

Species profiles, distribution maps and keys

Species profiles and distribution maps for each species are given in the first four Provisional Atlases (Edwards, 1997, 1998; Edwards & Telfer, 2001, 2002). Reference is made to the appropriate Provisional Atlas for each species in the Species accounts. Any updated distribution maps are available on the website of BWARS (www.bwars.com). The BWARS website also provides information sheets for the general public on *Dolichovespula media*, *Vespa crabro* and *V. velutina*.

The keys start with the assumption that an aculeate Hymenopteran specimen can be recognised. Recognition of the Aculeata within the order Hymenoptera and Hymenoptera within the class Insecta are left to other authors, e.g. Gauld & Bolton (1996). The following keys start by differentiating between the aculeate superfamilies and the families considered in this publication. Keys are then given to the genera and species. Species restricted to the Channel Islands, vagrants (e.g. *Eumenes papillarius*) and possible future species invasions (e.g. *Vespa velutina*) are also considered.

Key to Aculeata

1. Wings present 2

- Wings absent 7

2. Pronotum not laterally extending back to the tegulae .... **Apoidea**

- Pronotum laterally extending back to the tegulae (Fig. 9) ... **Chrysidoidea, Vespoidea** 3

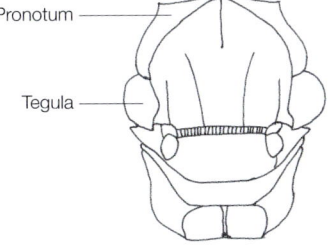

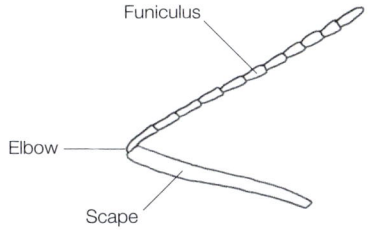

Figure 9. Dorsal view of the mesosoma. Figure 10. View of the antenna of an ant.

3. Antenna elbowed between the scape and the rest of the antenna (funiculus) (Fig.10) .... **Formicidae**

- Antenna not elbowed between the scape and the rest of the antenna (pedicel and flagellum) 4

4. Mesopleuron with a straight fine diagonal furrow to the mesopleural pit (Fig. 11). The hind tibiae reaching beyond the end of the gaster .... **Pompilidae**

- Mesopleuron without a straight diagonal furrow to the mesopleural pit (do not confuse with a pitted furrow found in mason and some chrysid wasps). The hind tibiae not reaching beyond the end of the gaster 5

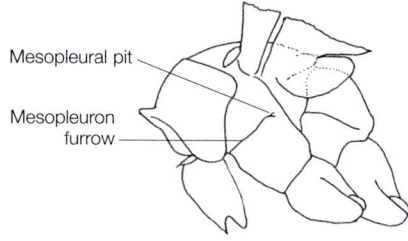

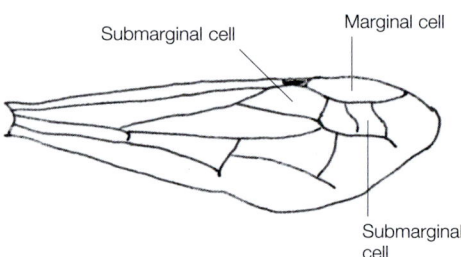

Figure 11. Lateral view the mesosoma of a pompilid.   Figure 12. View of the fore wing.

5. Fore wing with an open marginal cell and no complete submarginal cells. Hind wing without closed cells .... **Chrysidoidea**

- Fore wing with a closed marginal (except for female *Tiphia*) and submarginal cells (Fig. 12). Hind wing with at least two closed cells 6

6. Discal cell of the fore wing much longer than the subbasal cell (Fig. 13) ... **Vespidae** (p. 27)

- Discal cell of the fore wing usually shorter than the subbasal cell, at most of the same length (Fig. 14) .... **Tiphiidae, Mutillidae, Sapygidae, Scoliidae** (p. 13)

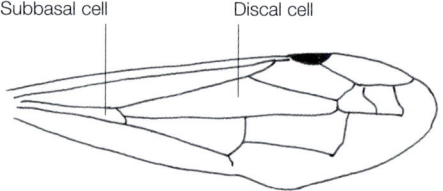

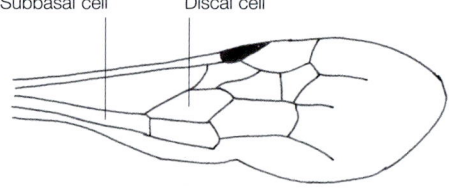

Figure 13. View of fore wing of a vespid.   Figure 14. View of fore wing of other vespoids.

7. Antenna elbowed between the scape and the rest of the antenna (funiculus) (Fig. 10) .... **Formicidae**

- Antenna not elbowed between the scape and the rest of the antenna (pedicel and flagellum) 8

8. Antenna consists of ten segments .... **Embolemidae, Dryinidae**

- Antenna consists of twelve or 13 segments 9

9. Head is flattened dorso-ventrally and the mouthparts are directed forwards. A carina or ridge is present between the antennae .... **Bethylidae**

- Head is not flattened and the mouthparts are directed ventrally. A carina or ridge is not present between the antennae .... **Tiphiidae, Mutillidae** (p. 13)

Note: The sexes of the five families considered in this handbook can be separated by the following characteristics: the female has an antenna of twelve segments and a gaster with six visible segments; the male has an antenna of 13 segments and a gaster with seven visible segments.

Further information can be found in Gauld & Bolton (1996).

Families Tiphiidae, Mutillidae, Sapygidae and Scoliidae

1. Wings absent (females of the genera *Methocha, Mutilla, Smicromyrme, Myrmosa*) 2

- Wings present 5

2. Mesosoma consists of three parts: prothorax, mesothorax, and metathorax with propodeum .... *Methocha* (p. 20)

- Mesosoma not of three parts 3

3. Dorsally mesosoma with a transverse suture. Ocelli present. Second gastral tergite without a felt line .... *Myrmosa* (p. 23)

- Dorsally mesosoma without a transverse suture. Ocelli absent. Second gastral tergite with a lateral felt line (Fig. 15) 4

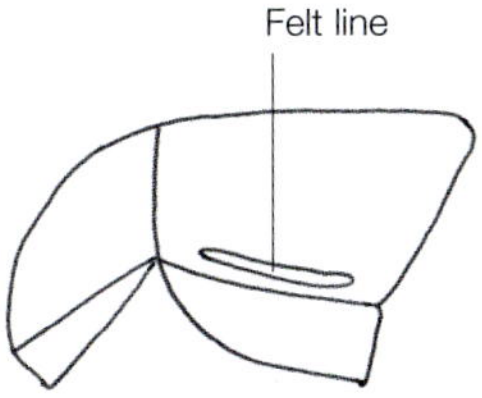

Felt line

Dorsal view of tergite six of *Smicromyrme*

Figure 15. Lateral view of second gastral tergite of *Mutilla* and *Smicromyrme*.

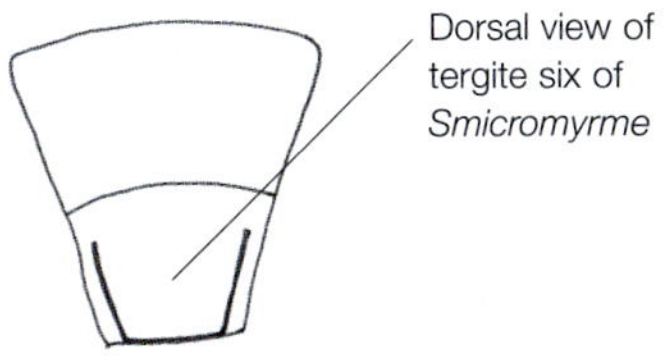

Figure 16. Dorsal view of tergite six of *Smicromyrme*.

4. First gastral tergite strongly convergent anteriorly becoming anteriorly less than half the width of the second gastral tergite. Gastral tergite six with a pygidial area (Fig. 16) .... *Smicromyrme* (p. 22)

- First gastral tergite weakly convergent anteriorly becoming anteriorly more than half the width of the second gastral tergites. Gastral tergite six lacks a pygidial area .... *Mutilla* (p. 21)

Two keys are provided for the winged sex or species to give confidence in arriving at the correct taxon. Use both keys. (See p. 13 for separating sexes.)

First key

5. Fore wing with two submarginal cells (Fig. 17) 6

\- Fore wing with three submarginal cells (Fig. 18) 8

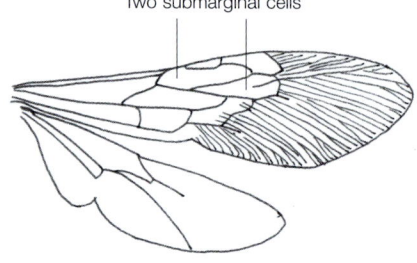

Figure 17. View of fore and hind wings of *Scolia*.

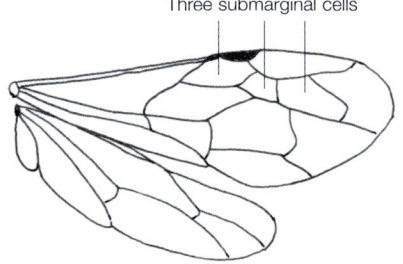

Figure 18. View of fore and hind wings of a sapygid.

6. Wings apically finely longitudinally wrinkled (Fig. 19). Dorsal and vertical surfaces of the propodeum divided into thirds by two longitudinal grooves. Male with three spines on sternite seven .... ***Scolia*** (p. 26)

\- Wings apically not wrinkled. Propodeum different. Male with a single spine on sternite seven 7

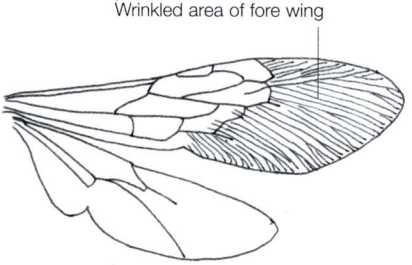

Figure 19. View of fore and hind wings of *Scolia*.

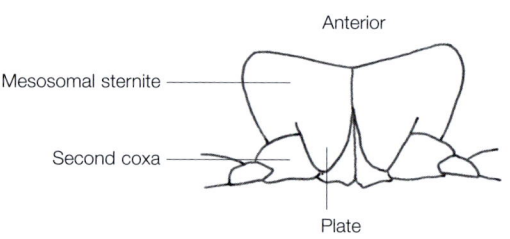

Figure 20. View of mesosomal sternite with plate extension of *Tiphia*.

7. Males and females. Middle coxae are widely separated and partially covered by plates at their base which are extensions of the mesosomal sternite (Fig. 20) .... ***Tiphia*** (p. 18)

\- Males only. Middle coxae are close together and not covered by plates. Just in front of each middle coxa there is an inward projecting spine .... ***Methocha*** (p. 20)

8. Gastral segment one not separated from segment two ventrally by a constriction or deep groove .... ***Sapygidae*** (p. 24)

\- Gastral segment one separated from segment two ventrally by a constriction or deep groove 9

9. Eyes not emarginated on their inner side (Fig. 22). No felt line present laterally on tergite two. Hind wing with a jugal lobe (Fig. 21) .... ***Myrmosa*** (p. 23)

- Eyes emarginated on their inner side (Fig. 24). Felt line present laterally on tergite two (Fig. 23). Hind wing without a jugal lobe (Fig. 25) 10

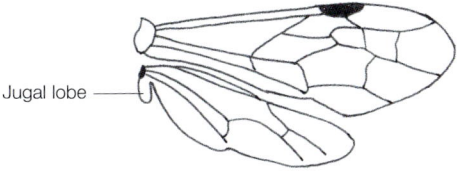

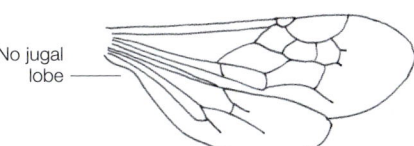

Figure 21. View of fore and hind wings of *Myrmosa*.

Figure 22. Frontal view of head of *Myrmosa*.

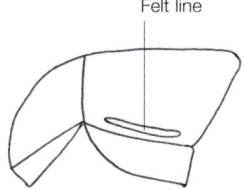

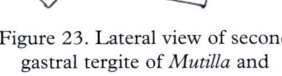

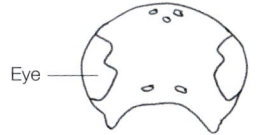

Figure 23. Lateral view of second gastral tergite of *Mutilla* and *Smicromyrme*.

Figure 24. Frontal view of head of *Mutilla* and *Smicromyrme*.

Figure 25. View of fore and hind wings of *Mutilla*.

10. First gastral tergite weakly convergent anteriorly becoming anteriorly more than half the width of the second gastral tergite. Vein Rs of fore wing reaches vein R+Sc near to the pterostigma; distance from where vein Rs meets vein R+Sc to the pterostigma shorter, or similar, in length to the pterostigma (Fig. 26) .... ***Mutilla*** (p. 21)

- First gastral tergite strongly convergent anteriorly becoming anteriorly less than half the width of the second gastral tergite. Vein Rs of fore wing reaches vein R+Sc far from the pterostigma; distance from where vein Rs meets vein R+Sc to the pterostigma longer than the pterostigma (Fig. 27) .... ***Smicromyrme*** (p. 22)

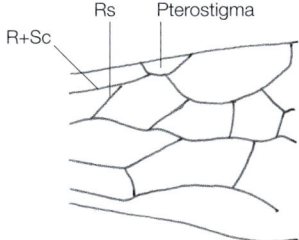

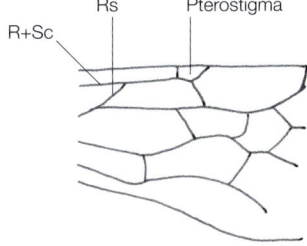

Figure 26. Part view of fore wing of *Mutilla*.

Figure 27. Part view of fore wing of *Smicromyrme*.

Second key

5. Gastral segment one not separated from segment two ventrally by a constriction or deep groove. Tergites one and two laterally with a fine ridge or carina .... ***Sapygidae*** (p. 24)

- Gastral segment one separated from segment two ventrally by a constriction or deep groove. Tergites one and two without a fine ridge or carina 6

6. Hind wing without a jugal lobe (Fig. 28) 7

- Hind wing with a jugal lobe (Fig. 29) 8

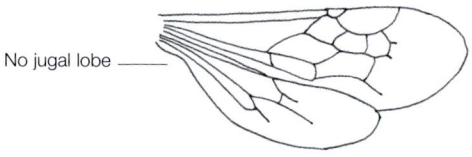

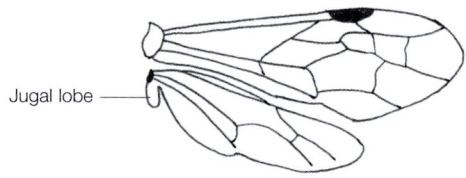

Figure 28. View of fore and hind wings of *Mutilla*.

Figure 29. View of fore and hind wings of *Myrmosa*.

7. First tergite weakly convergent anteriorly becoming anteriorly more than half the width of the second gastral tergite. Vein Rs of fore wing reaches vein R+Sc near to the pterostigma; distance from where vein Rs meets vein R+Sc to the pterostigma shorter, or similar, in length to the pterostigma (Fig. 30) .... **Mutilla** (p. 21)

- First gastral tergite strongly convergent anteriorly becoming anteriorly less than half the width of the second gastral tergite. Vein Rs of fore wing reaches vein R+Sc far from the pterostigma; distance from where vein Rs meets vein R+Sc to the pterostigma longer than the pterostigma (Fig. 31) .... **Smicromyrme** (p. 22)

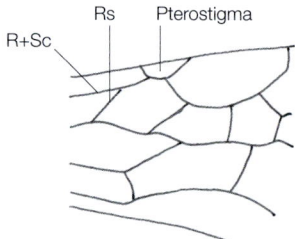

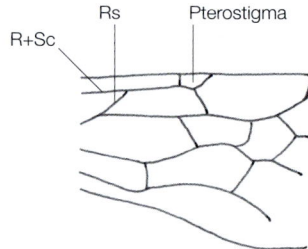

Figure 30. Part view of fore wing of *Mutilla*.

Figure 31. Part view of fore wing of *Smicromyrme*.

8. Wings apically finely longitudinally wrinkled (Fig. 32). Eyes emarginate on their inner side (Fig. 33) .... **Scolia** (p. 26)

- Wings apically not finely longitudinally wrinkled. Eyes not emarginate on their inner side 9

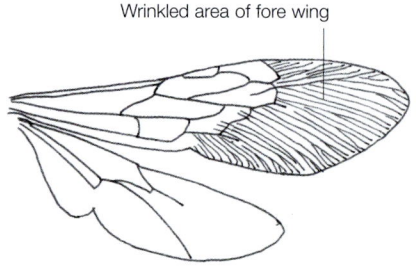

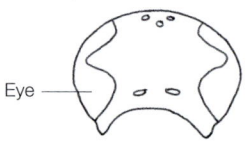

Figure 33. Frontal view of head of *Scolia*.

Figure 32. View of fore and hind wings of *Scolia*.

9. Males and females. Middle coxae widely separated and covered by plates which are extensions of the mesosomal sternite (Fig. 34) .... ***Tiphia*** (p. 18)

\-   Males only. Middle coxae close together and not covered by plates which are extensions of mesosomal sternite 10

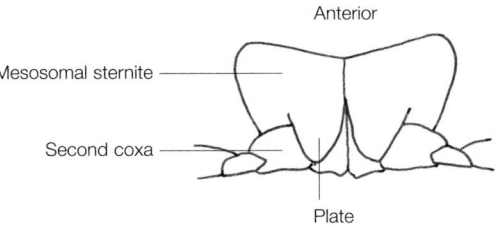

Figure 34. View of mesosomal sternite with plate extension of *Tiphia*.

10. Fore wing with two submarginal cells (Fig. 35) .... ***Methocha*** (p. 20)

\-   Fore wing with three submarginal cells (Fig. 36) .... ***Myrmosa*** (p. 23)

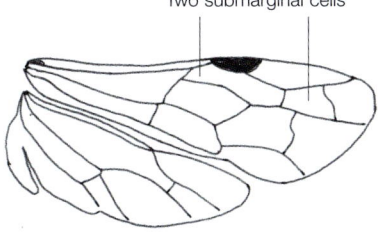

Figure 35. View of fore and hind wings of *Methocha*.

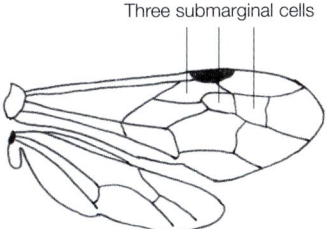

Figure 36. View of fore and hind wings of *Myrmosa*.

To some extent, a key is not necessary to separate these taxa because each, except for ***Myrmosa***, has one or two diagnostic characters.

Tiphia. Middle coxae are widely separated and partially covered by plates which are extensions of mesosomal sternite.

Methocha. With a pair of inward projecting spines just in front of the middle coxae.

Mutilla and ***Smicromyrme***. Felt line present laterally on tergite two.

Sapygidae. Gastral segment one not separated from segment two ventrally by a constriction or deep groove. Sides of tergites 1 and 2 with a fine ridge or carina.

Scolia. Wings apically finely longitudinally wrinkled. The dorsal and vertical surfaces of the propodeum divided into thirds by two longitudinal grooves.

Family Tiphiidae

A cosmopolitan, but mainly tropical, family of about 1500 species in seven subfamilies. Two subfamilies found in Britain: Tiphiinae and Methochinae.

Subfamily Tiphiinae

A cosmopolitan genus with about 300 species in nine genera. Two British species in one genus.

Genus **Tiphia** Fabricius, 1775

The British species are parasitoids on scarabaeid beetle larvae (*Aphodius, Amphimallon*). The female burrows into the soil to find a usually mature larval host in its cell. The wasp burrows below the host's cell before breaking into it, where she stings the larva and kneads it with her mandibles. An egg is laid usually on the lateral or ventral surface of the host in a fold of the cuticle. The paralysis is temporary, lasting 20-40 minutes, after which the larva becomes active, usually sufficiently so to continue feeding on grass roots. The larva takes about three weeks to eat its host. Pupation takes place in the host's cell. Probably one generation a year.

Diagnosis for British species

Females and males winged.

Eyes hairless.

Large tegula covering sclerites at the base of the fore wing.

Mesosternum expanded posterio-medially as a pair of small plates overlying the bases of the widely separated mid-coxae.

Dorsal surface of propodeum rectangular, bounded by keels and with three medial keels.

Posterior surface of propodeum triangular bounded posteriorly by a keel and with a medial keel.

Fore wing with two submarginal cells, the marginal cell is open in the female (Fig. 37).

Hind wing with jugal lobe, longer than the subbasal cell (Fig. 37).

Fore and mid tibiae with one spur, hind tibia with two spurs.

Mid and hind tibiae in female with stout spines.

Claws of tarsus bifid or dentate.

First gastral segment separated from the second by a constriction, particularly ventrally.

Male with an upwardly directed spine on the seventh gastral sternite.

Note: Preparation of specimens for identification. The propodeum should be clearly made visible by depressing the gaster.

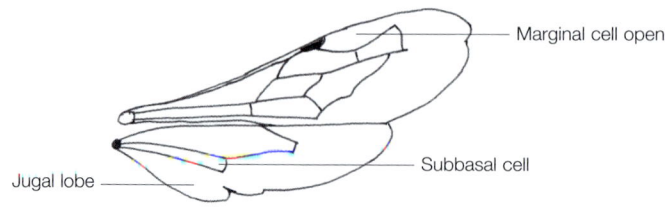

Figure 37. View of fore and hind wing of *Tiphia*.

Key to species

1. The propodeum laterally with prominent parallel-sided ridges and depressions with a shiny appearance. Gastral tergite one with distinct punctation towards the posterior margin except for the smallest specimens. Male with ridges in furrow at base of second gastral tergite strong and complete. Female with the mid and hind femurs and tibiae red. Generally larger, body length of female 7-14 mm, male 5-11 mm .... *femorata*

- The propodeum laterally with less prominent parallel-sided ridges which may be incomplete and cross ridges may be present. Gastral tergite one without distinct punctation towards the posterior margin. Male with ridges in furrow at base of second gastral tergite weak, sometimes incomplete and centrally missing. Female with the mid and hind femurs and tibiae black. Generally smaller, body length of female 5-7 mm, male 4-6 mm .... *minuta*

It is advised that named specimens of the two species be examined to fully appreciate the differences.

Species accounts

Tiphia femorata Fabricius 1775 Plates 11 & 12

Male black and female black with middle and hind trochanters, femurs and tibiae red. Length: females 7-14 mm, males 5-11 mm. Distributed from Cornwall to Kent, south Wales and north to Norfolk. Also recorded from the Channel Islands. Distribution map in Edwards & Telfer (2001). Elsewhere, found in much of Europe and also recorded from far eastern Russia. Found in open, sunny, often disturbed places such as coastal sand dunes (fore and middle dunes), inland heaths, calcareous grasslands, undercliffs and landslip areas. Also recorded from a clearing with basic grassland in a coniferous plantation. Probably single-brooded. Flight period most likely during July and especially August, but also much less frequently during June and September. Found to visit the following flowers: angelica, carrot, fennel, hogweed, ragwort, rock samphire, spurge and wild parsnip. Smith (1858) found that the males passed the nights enclosed in the petals of flowers.

Tiphia minuta Van der Linden, 1827 Plate 13

Females and males black. Length: females 5-7 mm, males 4-6 mm. Widespread in Britain, south of the Humber. There are isolated records from Westmorland, Yorkshire, Isle of Man, Ayrshire and Mid Perthshire. There is also one record from Ireland (Wicklow). Distribution map in Edwards & Telfer (2001). Falk (1991) listed this species as Nationally Scarce (Nb). Elsewhere, found in many parts of Europe and also in far eastern Russia. Found on heathland, downland and other types of grassland, open woodland and coastal dunes. Flight period from May until August with a peak in June and July. Found to visit the following flowers: wild carrot, garden chervil, evening-primrose, ground-elder and wild parsnip.

Subfamily Methochinae

A worldwide distributed family, although absent from the Australian region, with about 45 species in two genera. One British species.

Genus **Methocha** Latreille, 1804

The larvae are parasitoids on larvae of tiger beetles (*Cicindela*). The female runs over the surface of the ground looking for the burrow of its host. When found, the wasp allows the host's larva to grasp her around her heavily-armoured thorax. As the host comes out of its burrow the wasp bends its gaster down and stings the larva below the head capsule. The larva is further stung several times and quickly becomes immobilised. The wasp then pulls its prey deep into the burrow where she lays a single egg on the ventral side, usually behind the coxae of the hind legs. The wasp fills the burrow with grains of sand, small twigs and small fibrous pieces of humus before leaving. Probably one generation a year.

Diagnosis of British species

Female wingless, male winged.

Eyes sparsely hairy.

Female with mesosoma divided into three parts: prothorax, mesothorax, and metathorax with propodeum.

Female with mesothorax narrower than the prothorax and metathorax with propodeum.

Male with short tegula not concealing sclerites at base of fore wing.

Mesopleuron of male with an epicnemial ridge and a large central hairy depression.

Male with the metasternum expanded posterio-medially as a pair of small spines.

Mid coxae close together.

Propodeum of male rounded and latticed (clathrate).

Fore wing of male with two submarginal cells and marginal cell long and pointed.

Hind wing of male with jugal lobe, shorter than half the length of the subbasal cell.

Female with one tibial spur on each leg. Male with one tibial spur on fore leg and two tibial spurs on mid and hind legs.

Claws of tarsus dentate in male, simple in female.

First gastral segment separated from the second by a constriction, particularly ventrally.

Male with a strong upwardly directed spine on the seventh sternite.

Species account

Methocha articulata Latreille, 1792 Plates 13 & 14

This species was previously known as *Methocha ichneumonides* Latreille 1804.

Male black and female black with part of the antenna and mesosoma red. Length: females 3-9 mm, males 7-11 mm. Distributed from south Wales and England north to Yorkshire (Allerthorpe and Strensall Commons) and Cumberland (Skirwith). Distribution map in Edwards (1998). Falk (1991) listed this species as Nationally Scarce (Nb). Elsewhere, found in many parts of Europe and North Africa. Found on open sandy areas in warm sunny situations both on the coast (sand dunes and clay cliffs) and inland (heathlands and sand pits). Adults active during June, July and August, and rarely during May and September. Males have been found on umbellifers and thistles.

Family Mutillidae

A cosmopolitan distribution, although a mainly tropical, family of about 5000 species in seven subfamilies. Two subfamilies found in Britain: Mutillinae and Myrmosinae.

Subfamily Mutillinae

A cosmopolitan distribution, although richest in the Old World and absent from most of Australia, consisting of about 1800 species in about 40 genera. Two genera found in Britain, *Mutilla* and *Smicromyrme*.

Diagnosis of British species

Females without wings, males with wings.

Ocelli absent.

Female coarsely and closely punctured with conspicuous patches of silvery pubescence. Male less coarsely punctured but with patches of silvery pubescence.

Female mesosoma box-like, lacking dorsal sutures.

Tegulae of male very large.

Fore wing of male with three submarginal cells.

Fore wing of male with translucent pterostigma.

Hind wing of male without a jugal cell.

Fore tibia with one spur, mid and hind tibiae with two spurs each.

Middle coxae of female close together but clearly separated in the male.

Second gastral tergite with a secretory felt line on each side.

Genus **Mutilla** Linnaeus, 1758

Distributed worldwide except the Americas, with many species. One British species.

The British species is a parasitoid of bumblebees and occasionally honeybees. The female enters the nest of the host and lays an egg in a cocoon cell with a prepupa or young pupa. The parasitoid larva eats the host and then spins a cocoon within the host's cell. The adult, on emergence, feeds on the honey stores of the host. Females overwinter as adults but males leave the host nest but do not survive the winter. The female may overwinter in the host's nest or elsewhere, e.g. at the roots of low herbage on chalk grassland.

Species account

Mutilla europaea Linnaeus, 1758 Plates 1 & 2

Females often called 'velvet ants' due to the lack of wings and ant-like appearance. Females black with mesosoma mainly red. Males black with dorsal mesosoma mainly red. Length: females 7-15 mm, males 9-14 mm. The British distribution is largely disjunct, being widespread in southern England, sporadically in north-eastern England and central Scotland. The species may be decreasing. Falk (1991) listed this species as Nationally Scarce (Nb). Distribution map in Edwards (1997). Elsewhere, found throughout Europe and the Palaearctic. Associated with heathlands, moorlands, chalk grasslands and woodlands. Female adults active mainly from June until September, but particularly during August, rarely during February, March, April, May and October. Male adults active mainly during July and August, sometimes during September and rarely during June and October. Females have been associated with the flowers of yarrow and parsnip and males with wild carrot, angelica, yarrow and scabious. Adults have also been found on flowers of bramble and fleabane.

Genus **Smicromyrme** Thomson, 1870

Distributed throughout the Oriental, Palaearctic and Afrotropical regions consisting of a moderate number of species. One British species.

The British species is a parasitoid of a wide variety of ground nesting wasps and bees including wasps of Crabronidae and Pompilidae and bees of Halictidae. The female enters the burrow of its host and bites open a cell. If the host's larva is immature she closes the cell and leaves. However, if a mature larva is present she inserts her gaster into the cell and lays an egg on the host which is not stung. The wasp larva is an ectoparasitoid on the larva or pupa of the host. On reaching maturity, the parasitoid spins a cocoon within that of the host.

Species account

Smicromyrme rufipes (Fabricius, 1787) Plates 3 & 4

Female black with lower part of the head, basal part of antenna, mesosoma and legs red. Male black except for lateral pronotum, mesoscutum, mesoscutellum and tegulae red. Length: females 3-6 mm, males 4-7 mm. In England distributed from Dorset to Kent, including the Isle of Wight, and north to Oxfordshire, Bedfordshire, Cambridgeshire and Norfolk. Also recorded from the Channel Islands. Falk (1991) listed this species as Nationally Scarce (Nb). Distribution map in Edwards (1998). Elsewhere, found in many parts of Europe. Associated with open sandy habitats in warm sunny situations both on the coast, *e.g.* sand dunes, and inland, *e.g.* heathland and sand pits. Adults are mainly active during July and August, sometimes during June, and rarely during May and September. Males have been found on the flowers of umbellifers and ragworts.

Subfamily Myrmosinae

Distributed throughout the Holarctic and Oriental regions consisting of less than 50 species in about eight genera. One British species.

Genus **Myrmosa** Latreille, 1796

The genus is Holarctic in distribution with a small number of species.

The British species is a parasite, probably a parasitoid, on ground nesting wasps of Crabronidae and bees of Halictidae.

Diagnosis of British species

Females without wings, males with wings.

Ocelli present.

Female coarsely and closely punctured, hairy with hairs forming fringes on gastral tergites, particularly tergites two to five. Male less coarsely punctured but with patches of silvery pubescence.

Female mesosoma box-like, divided into two by a suture.

Tegulae of male covering the base of the fore wings.

Fore wing of male with three submarginal cells.

Hind wing of male with a short jugal cell, less than half the length of the subbasal cell.

Middle coxae close together.

Fore tibia with one spur and mid and hind tibiae with two spurs each.

Second gastral tergite without a secretory felt line on each side.

Species account

Myrmosa atra Panzer, 1801 Plates 5 & 6

Females red with most of head, mandibles, upper part of antenna, posterior part of gastral tergite and sternite two and following tergites and sternites mainly black. Male black. Length: females 4-7 mm, males 5-11 mm. Distributed throughout England and Wales including the Isle of Man and the Channel Islands. Widely distributed in southern Ireland and County Tipperary (Stelfox, 1933). Distribution map in Edwards (1998). Elsewhere, found in many parts of Europe and Asia (Iran, Siberia). Associated with habitats with sandy areas (*e.g.* lowland heathland), chalky banks, slopes in woods and roadsides. Adults are mainly active during July and August, sometimes during June and September, and rarely during May and October. Male have been found visiting flowers of bramble and umbellifers such as wild carrot and wild parsnip.

Family Sapygidae

This family has a worldwide distribution, except for the Australian region, consisting of about 80 species in two subfamilies. One British subfamily.

Subfamily Sapyginae

Consists of several species in a few genera. Two British genera, *Sapypa* and *Monosapyga*.

Diagnosis of British species

Both sexes similar and winged.
Large eyes with inner borders deeply emarginated.
Fore wing with three submarginal cells.
Hind wing with a small jugal lobe, shorter than the subbasal cell.
Middle coxae close together.
Fore tibia with one spur, mid and hind tibiae with two spurs each.
Gastral segment one and two not separated ventrally by a constriction.
Sides of gastral tergite one and two with a fine ridge or carina.
Female with the gastral sternite six projecting beyond gastral tergite six.

> Note: Preparation of specimens for identification. The antennae should be spread away from the body.

Key to species

1. Females with gastral tergites two and three red and gastral sternites often lacking yellow markings. Male with antennal segment 13 very short, less than half the length of antennal segment twelve and segment twelve of similar size to other apical segments. Segments twelve and 13 without a ventral depression .... ***Sapyga quinquepunctata***

- Females with gastral tergites two and three and some gastral sternites with yellow spots. Male with antennal segment 13 long and of similar size to segment twelve and both segment twelve and 13 swollen in comparison with the other antennal segments and each with a ventral depression .... ***Monosapyga clavicornis***

Genus **Monosapyga** Pic, 1920

Almost entirely Holarctic in distribution with one British species.

The British species is a cleptoparasite of the bees *Chelostoma florisomne* (L.) and species of *Osmia*. The female introduces her egg into the cell of the host with the aid of the ovipositor, which is not exclusively used as a sting. On hatching, the first instar larva has large mandibles and despite being without legs, is active and proceeds to eat the host's egg. Then it moults into the next larval instar which has smaller mandibles, and proceeds to feed on the host's provisions. Probably one generation a year.

Species account

Monosapyga clavicornis (Linnaeus, 1758) Plates 7 & 8

Female black with yellow markings on the clypeus, frons, tip of last antennal segment, pronotum, tibiae and gastral tergites and sternites. Male black with the clypeus mainly yellow and yellow markings on the ventral surface of the antennae, tibiae and gastral tergites and sometimes pronotum and gastral sternites. Length: females 7-10 mm, males 6-10 mm. Distributed in England and Wales from Cornwall to Kent and north to Yorkshire. Distribution map in Edwards (1998). Falk (1991) listed this species as Scarce (Nb). Elsewhere, found in many parts of Europe, North Africa and Asia (Turkey, Caucasus). Associated with dead wood such as posts, old trees and stumps in sunny situations which are the nesting sites of its hosts. As such, found in a wide range of habitats including lowland heaths and sandy places, river margins, limestone grassland, clay woodland, fenland, parkland, coastal areas and gardens. Adults are mainly active during June, and sometimes during May and July.

Genus **Sapyga** Latreille, 1796

Almost entirely Holarctic in distribution with one British species.

The British species is a cleptoparasite on bees of the genera *Osmia* (*O. bicornis* (L.), *O. leaiana* (Kirby), *O. aurulenta* (Panzer), *O. caerulescens* (Linn.)) and *Chelostoma*. The female enters the nest of its host and lays an egg on or near the egg of its host. Gauld & Bolton (1996) indicate that the sting is used to penetrate the cell wall, thus acting as an ovipositor. P. Westrich (pers. comm., 1999) says that the egg is placed directly in the cell. On hatching, the first instar larva, which has large mandibles, eats the egg of the host. Later instars have smaller mandibles and feed on the stored food of the host. One generation a year.

Species account

Sapyga quinquepunctata (Fabricius, 1781) Plates 9 & 10

Female black with the first and second gastral tergite and sternite mainly red, and with yellow marking on the frons and the fourth, fifth and sixth gastral tergites and sometime the clypeus, ocular sinus, pronotum and tibiae. Male black with the ventral surface of the antennae reddish, the clypeus mainly yellow and yellow marks on the ocular sinus, and gastral tergites and sternites and sometimes the pronotum and tibiae. Length: females 9-13 mm, males 7-11 mm. Distributed in England and Wales from Cornwall to Kent and north to Cumbria and Yorkshire. Distribution map in Edwards & Telfer (2001). Elsewhere, found in many parts of Europe, the Middle East (Syria, Jordan, Israel, Turkey) and North Africa. Associated with a wide range of habitats: woodland and scrubland, lowland heaths, chalk quarries, coastal cliffs, gardens and cemeteries. Often found flying around the nest sites of its host: mud walls, holes in mortar, dead wood, old wooden posts and snail shells. Males are active during May and June and rarely during July. Females are active mainly during June, often during May and July, or more rarely during April and August. It has been recorded visiting the flowers of thyme.

Family Scoliidae

A cosmopolitan, but predominantly tropical, family of about 300 species in two subfamilies. One Channel Islands subfamily. The other subfamily consists of two species.

Subfamily Scoliinae

Several genera with one Channel Islands genus.

Genus **Scolia** Fabricius, 1775

This species is a parasite of beetle larvae, usually Scarabaeidae. The female flies very close to the surface of the ground, landing above the habitat of its host. The female tunnels in the soil looking for a large host larva. On detecting a larva, the female stings, completely paralysing it, lays an egg on it and departs. The scoliid larva is an ectoparasitoid and completes its development within the cell of its host.

Diagnosis of British species

Both sexes are fully winged.

Inner border of eyes deeply emarginated.

The upper and vertical surfaces of the propodeum divided into thirds by two longitudinal grooves.

Fore wing with two submarginal cells.

Hind wing with a long jugal lobe, longer than the subbasal cell.

Distal portions of all wings finely and longitudinally striolate giving a dense corrugated appearance.

Fore coxae near each other, middle and hind coxae widely separated.

Female with middle and hind tibiae stout and heavily spined.

First and second gastral sternites separated from a deep constriction.

Last visible gastral sternite of male armed with three spines.

Species account

Scolia sexmaculata Müller, 1775.

The species is black with yellow marks on the second and third gastral tergites. Body and legs very hairy. Length: females 10-14 mm, males 8-14 mm. Restricted to the Channel Islands (Jersey). Elsewhere, found in many parts of Europe (particularly in southern parts including Crete), North Africa and Asia (Turkey). Adults are active during July and August. It has been associated with the flowers of sea-holly, wild marjoram, mint and tamarisk.

Family Vespidae

Worldwide distribution but predominantly in tropical regions. In excess of 4700 species in six subfamilies. Three subfamilies in Britain: Eumeninae, Polistinae and Vespinae.

Diagnosis for British species

Females and males winged.

Eye with inner margin deeply emarginated.

Fore wings longitudinally folded when at rest.

Discal cell of fore wing long, longer than the subbasal cell.

Apical margin of subdiscoidal cell (AMDC) of fore wing straight or slightly curved (Fig. 38).

Hind wing with either a short jugal lobe which is less than one third of the length of the subbasal cell (Fig. 39) or absent.

Mid and hind coxae touching each other.

Mid femur with a basal ring.

First gastral tergite partially fused at base to first sternite and first tergite overlapping first sternite.

First gastral sternite separated from second sternite by a deep constriction.

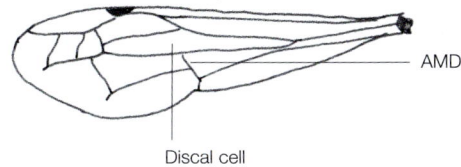

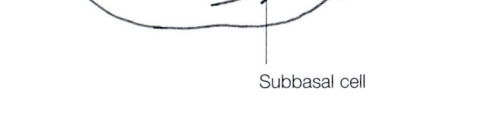

Figure 38. View of fore wing of a vespid.   Figure 39. View of hind wing of a vespid.

Key to subfamilies

1. Mid-tibia with one spur. Mandibles elongate and crossing over when fully closed. Tarsal claws bifid. Hind wing with a jugal lobe. Mesoscutum with a posterio-lateral lobe projecting backwards (the parategula, Fig. 40) .... **Eumeninae** (p. 28)

- Mid-tibia with two spurs. Mandibles short and only slightly overlapping when fully closed. Tarsal claws simple, not bifid. Hind wing with or without a jugal lobe. Mesoscutum without parategulae 2

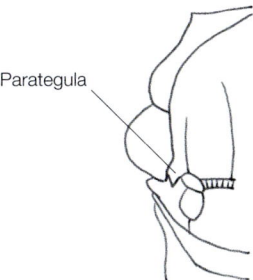

Figure 40. View of left part of the mesosoma of a eumenine.

2. Dorsal surface of the first gastral tergite rounded with the anterior and dorsal surfaces merging together (Fig. 41). Hind wing with a jugal lobe .... **Polistinae** (p. 47)

- Dorsal surface of the first gastral tergite truncate with the anterior vertical surface almost at right angles to the dorsal surface (Fig. 42). Hind wing without a jugal lobe .... **Vespinae** (p. 50)

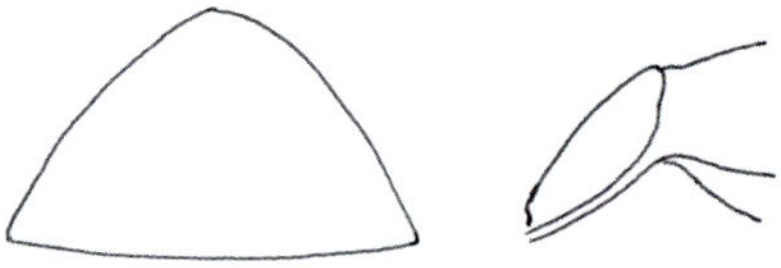

Figure 41. Dorsal and lateral view of the first gastral tergite of a polistine.

Figure 42. Dorsal and lateral view of the first gastral tergite of a vespine.

Subfamily Eumeninae

Worldwide in distribution with about 3500 species in over 200 genera. Nine British genera with 23 species (including 1 species restricted to the Channel Islands and, in addition, 2 vagrant species).

Usually solitary, narrow-bodied wasps. Black with yellow or whitish-yellow bands and spots. The size and colour given for each species should be treated with caution. The British species are all solitary with each female building its own cells entirely or partially of mud. Nests may be in excavated burrows underground, in hollowed-out plant stems, clay nests attached to plant stems or a hard substrate or built in pre-existing cavities. Adults are predatory, hunting the larvae of moths and beetles to feed their offspring. The female searches for a nest site, building a nest of a few cells and provisioning each cell with prey. One egg is laid at the end of a thread which is attached to the top of the cell before provisioning starts. After each cell is provisioned with several prey items it is sealed. Overwinter in the prepupal stage in the cell. One or more generations a year. Parasitised by chrysidid wasps. Spradbery (1973) provides a good introduction to the eumenines.

Diagnosis of British species

All species are solitary.

Mandible elongate, crossing over when fully closed.

Short jugal lobe present on the hind wing.

Middle tibia with a single spur.

Claws bifid.

Key to British genera

Note: A specimen should be pinned through one side of the mesoscutum so that the centre and one side of the mesoscutum are not damaged.

The mandibles should be opened and not obscured by the labrum.

The antennae should be positioned away from the head.

The gaster should hang downwards so that the propodeum is clearly visible.

The segments of the gaster should be restored to normal position (if they have become telescoped) by pulling out with fine forceps.

Make sure the sides of the thorax and propodeum are visible by letting the legs hang downwards.

Make sure the second gastral sternite is visible from a ventral aspect despite the need for the gaster to hang downwards.

There is no need to extract the sting or the male genitalia.

1. First gastral tergite petiolated so that it is about half as wide as the second gastral tergite (Fig. 43) .... ***Eumenes*** (p. 32)

- First gastral tergite non-petiolated, and nearly as wide as the second gastral tergite (Fig. 44) 2

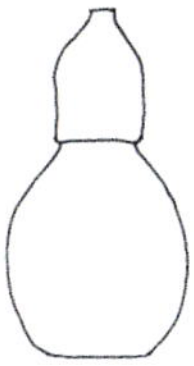

Figure 43. Dorsal view of first and second gastral tergites of *Eumenes*.

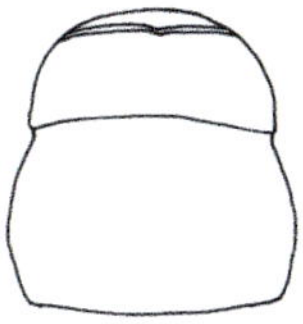

Figure 44. Dorsal view of first and second gastral tergites on a non- *Eumenes*

2. The first gastral tergite with a transverse ridge just posterior to the anterior face (Fig. 44) 3

- The first gastral tergite without a transverse ridge just posterior to the anterior face 4

3. First gastral tergite with a central longitudinal linear depression (Fig. 45) .... ***Symmorphus*** (p. 33)

- First gastral tergite without a central longitudinal linear depression .... ***Ancistrocerus*** (p. 35)

Figure 45. Dorsal view of first and second gastral tergites of *Symmorphus*.

4. Tegula pointed posteriorly (Fig. 46A) 5

- Tegula rounded posteriorly (Fig. 46B) 7

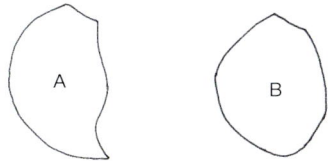

Figure 46. View of left tegulae.

5. Thorax and propodeum elongate, each being nearly twice as long as wide. The large punctures on the mesoscutum are separated by flat areas covered in micropunctures. No red marks present on the first gastral tergite. Mandible with four teeth in both sexes (Fig. 47) .... *Microdynerus* (p. 41)

- Thorax and propodeum each being only a little longer than wide. The large punctures on the mesoscutum are separated by ridges, either with or without micropunctures. Red mark may or may not be present on the first gastral tergite. Mandible with a different arrangement of teeth from *Microdynerus* in both sexes 6

Figure 47. View of mandible of *Microdynerus*.

6. First gastral tergite with red marks. Head of female with a depressed area behind the ocelli extending laterally to the lateral margins of the posterior ocelli. Male with the anterior margin of the clypeus with a deep semi-circular notch (Fig. 48). Mandible of the female with five distinct teeth Fig 49). Mandible of the male with three teeth, the second tooth apically truncate, and a deep rectangular notch between the second and third teeth (Fig. 50) .... *Pseudepipona* (p. 42)

- First gastral tergite without red marks. The head of female with a depressed area behind the ocelli extending laterally beyond the lateral margins of the posterior ocelli. Male with the anterior margin of clypeus shallowly notched, less than a half-circle (Fig. 51). Mandible in both sexes with four teeth, but the fourth tooth is double pointed (Fig. 52) .... *Euodynerus* (p. 42)

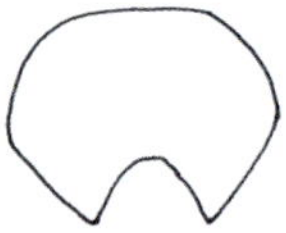

Figure 48. View of male clypeus of *Pseudepipona*.

Figure 49. View of female mandible of *Pseudepipona*.

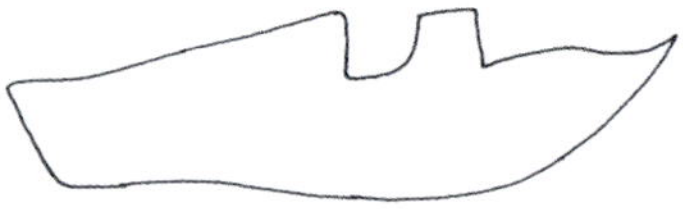

Figure 50. View of male mandible of *Pseudepipona*.

Figure 51. View of male clypeus of *Euodynerus*.

Figure 52. View of mandible of *Euodynerus*.

7. Apical segments of labial palps with long conspicuous bristles (Fig. 53) ....
.... ***Pterocheilus*** (p. 43)

- Apical segments of labial palps without long conspicuous bristles 8

Figure 53. Frontal head view of *Pterocheilus*.

8. Head much widened and bulbous behind the eyes (Fig. 54). On the mesoscutum there are no ridges between the closely-set punctures. Female with a deep notch on the anterior margin of the clypeus (Fig. 55). Female with a deep rectangular notch between the third and fourth teeth of the mandible (Fig. 56). Male without a spine on the gena or the middle coxa, and without deep notches on the middle femur .... ***Gymnomerus*** (p. 43)

- Head less widened and bulbous behind the eyes (Fig. 57). On the mesoscutum ridges separate the closely-set punctures. Female with a shallow notch on the anterior margin of the clypeus (Fig. 58). Female without a deep rectangular notch between the third and fourth teeth of the mandible (Fig. 59). Male either with a spine on the gena and the middle coxa or with deep notches on the middle femur .... ***Odynerus*** (p. 44)

Figure 54. Dorsal view of
head of *Gymnomerus*.

Figure 55. View of female
clypeus of *Gymnomerus*.

Figure 56. View of female mandible
of *Gymnomerus*.

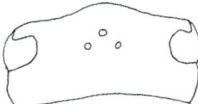

Figure 57. Dorsal view of
head of *Odynerus*

Figure 58. View of female
clypeus of *Odynerus*.

Figure 59. View of female
mandible of *Odynerus*.

Note: The vagrant, ***Rhynchium oculatum*** can be separated from all other species by the posterior part of the mesoscutum and scutellum being smooth and shining and almost without punctures. A male was observed on a stonecrop flower in a Nottingham garden (SK597457) (Archer, 1998a). The male specimen had a body length of 15 mm and was reddish brown in colour except for a yellow clypeus.

Genus **Eumenes** Latreille, 1796

This genus consists of about twelve species in Europe with one British species and a vagrant species.

Key to British species of *Eumenes*

1. Hairs of clypeus short and even. Erect hairs on the second gastral tergite longer and uneven (Fig. 60) .... ***papillarius***

- Hairs on clypeus not even being longer dorsally. Erect hairs on the second gastral tergite short and even (Fig. 61) .... ***coarctatus***

 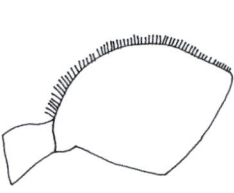 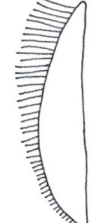 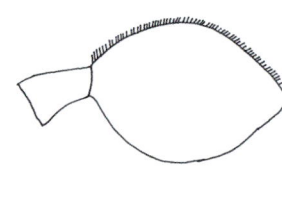

Figure 60. Lateral view of clypeus and second gastral tergite of *Eumenes papillarius*.

Figure 61. Lateral view of clypeus and second gastral tergite of *Eumenes coarctatus*.

Species accounts

Eumenes coarctatus (Linnaeus, 1758) Plates 32 & 33

Black with yellow markings on the head, mesosoma and gaster. Gastral tergites one to four often with an apical yellow band. Gastral tergite two with two large yellow spots. Length: female 13-15 mm, male 9-13 mm. Distributed from West Cornwall to East Sussex and north to Glamorgan, Wiltshire and Buckinghamshire. Distribution map in Edwards & Telfer (2001) and Archer (2003a). Falk (1991) listed this species as Nationally Scarce (Na). Elsewhere, found in many parts of Europe and in Asia east to Mongolia and far eastern Russia. Found on lowland heathlands with resources of clayey soil and water needed to form mud cells. Adults active usually from June until September, rarely during May and October. Males are most likely to be seen during June and July and females during July and August. The female builds clay pots (cells) which are attached to the stems of heather, or gorse, or on stones and walls. Several pots may be built side-by-side but still remain distinct (Lucas, 1931). The pots are made from clayey soil which is made into a soft paste by mixing with water. Each pot has a funnel-shaped constriction, usually called a neck, acting as a mouth which is so small that the female cannot enter the pot. The prey consists of small lepidopteran larvae such as tortricids and geometrids. The mud of the neck is used to seal the pot after it has been provisioned (Barrington & Edwards, 1991). Known to visit flowers of angelica, bramble, dwarf elder, heather and heaths. Spradbery (1973) quotes *Chrysis ignita* (L.) and several ichneumonid species from a nineteenth century German paper (Ratzeburg, 1852) as parasites. The ichneumonid, *Acroricnus stylator* (Thunberg) (as *Linoceras macrobatus* Bridgman) is reported as an ectoparasitoid. The braconid, *Aleiodes modestus* (Reinhard) (as *Rhogas modestus*), and the ichneumonid, *Dusona subreptus* Tst. (as *Campoplex subreptus*), have been reported as endoparasites on the prey (Bridgman, 1887).

Eumenes papillarius (Christ, 1791)

Black with yellow markings on the head, mesosoma and gaster with yellow markings more extensive than *E. coarctatus*. Gastral tergites one to five often with an apical yellow band. Gastral tergite two with two large yellow spots. Length: female 13-15 mm, male 9-13 mm. Only known from two records, both from Yorkshire (Guichard, 1991; Archer, 2000b), so regarded as a vagrant species. Elsewhere, found mainly in middle Europe, but also Spain and in Asia east to Turkmenistan and Kazakhstan. Builds mud cells which are attached to stones, crevices in walls and on window frames.

Genus **Symmorphus** Wesmael, 1836

This genus consists of ten species in Europe with four British species. They are tube-nesters using existing cavities such as plant stems, e.g. bramble and elder, vacated burrows of wood-boring insects, straws of thatched roofs, old walls and even vertical sandy banks. All the species seem to prefer damp places, often near streams and ditches. The mating flight path of the males and females is along a particular stretch of foliage, such as poplar, aspen and alder, in bright sunlight (Guichard, 1972). After mating, when a suitable nesting site has been found, the female clears away any remaining pith and debris. Next she collects clay which is softened with water stored in her crop. Often, a preliminary plug of clay is placed at the end of the burrow before the cells are started. The clay partitions between the cells have a smooth, convex inner surface and a rough outer surface, which probably helps the full-grown larva to orientate itself for its eventual escape as an adult from the cell. The cells are arranged linearly. An egg is laid in each cell before it is provisioned. The prey is immobilized by stinging and carried by the mandibles and forelegs. Prey items are packed tightly in the cell, avoiding the egg. Usually, the egg takes two to three days to hatch, with the larval stage taking one to two weeks probably with five instars. When feeding is completed the larva may rest for about a day and void its gut contents at the inner end of the cell before spinning a cocoon which is anchored at the base of the cell. When the species is univoltine, the prepupal stage lasts for six to eight months, the pupal stage for one week and the newly emerged adult rests in the cell for two to three weeks before emerging. When the species is multivoltine, the prepupal stage lasts about one week. Adults have been recorded visiting flowers of figwort, umbellifers (e.g. hogweed, wild parsnip) and spurge. Jørgensen (1942) gives detailed information about life histories of *S. bifasciatus* and *S. connexus*.

Key to the species of *Symmorphus*

1. Lateral anterior pronotal angles pointed and projecting forwards. Pronotum lacks an anterior transverse ridge (Fig. 62) .... ***gracilis*** (p. 35)

- Lateral anterior pronotal angles rounded, or if pointed, projecting laterally. Pronotum with an anterior transverse ridge (Fig. 63) 2

Figure 62. Dorsal view of pronotum of *Symmorphus gracilis*.

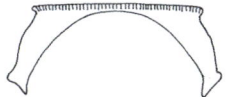

Figure 63. Dorsal view of pronotum of non- *Symmorphus gracilis* species.

2. The punctures on the area in front of the anterior ocellus separated by flat surfaces. (Hair characteristics as for *S. bifasciatus*) .... *connexus* (below)

- The punctures on the area in front of the anterior ocellus so close together that they are separated only by ridges 3

3. Hairs on the sloping face of the first gastral tergite long, as long as the hairs on the propodeum (left of Fig. 64). Generally hair on the head, thorax and propodeum long. The hair on the mesoscutum as long as the width of the antennal scape ... *crassicornis* (p. 35)

- Hairs on the sloping anterior face of the first gastral tergite shorter than the hairs on the propodeum (right of Fig. 64). Generally hair on the head, thorax and propodeum short. The hair on the mesoscutum shorter than the width of the antennal scape .... *bifasciatus* (below)

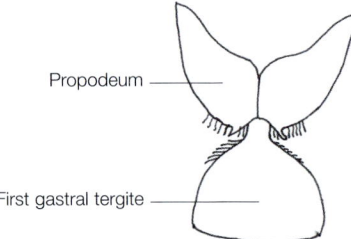

Figure 64. Dorsal view of propodeum and first gastral tergite of *Symmorphus crassicornis* (left side) and *S. bifasciatus* (right side).

Species accounts

Symmorphus bifasciatus (Linnaeus, 1761) Plates 45 & 46

This species was previously known as *S. mutinensis* (Baldeni, 1894), *S. sinuatus* (Fabricius, 1793) and *S. sinuatissimus* Richards, 1935. Black with yellow markings. Gastral tergites one, two and four usually with apical yellow markings. Length: female 7-10 mm, male 6-9 mm. Distributed throughout England, Wales and Ireland, including the Isle of Man and the Channel Islands, and Scotland north to the Highland region. Distribution map in Edwards (1997) and Archer (2003a). Elsewhere, found in middle and northern Europe and in Asia east to Siberia and Japan. Adults active usually from late June to early August, sometimes during early June and late August and rarely during May. Besides nesting in plant stems, it has been found nesting in disused plant galls of *Andricus kollari* (Hartig). Prey consists of larvae of the chrysomelid beetle, *Phrators vulgatissima*, L. Each cell is provisioned with 10-17 beetle larvae. The parasite, *Chrysis angustula*, Schenck has been associated with this species in Europe (Schneider, 1991), as well as *C. ignita* (L.) (Kunz, 1994).

Symmorphus connexus (Curtis, 1826)

Also previously known as *S. bifasciatus*. Black with yellow markings. Length: female 8-9 mm, male 6-8 mm. Gastral tergites one, two and four usually with apical yellow markings. Distributed from Dorset to Kent and northwards to Oxfordshire and Lincolnshire. Since 1970 it has undergone a significant decline. Recent records have been from Kent, Hampshire and Oxfordshire. Map of British distribution in Edwards (1997) and Archer (2003a). Listed as RDB3 in Shirt (1987) and Falk (1991). Overseas, found in northern and central Europe and east to central Asia. From the few records available adults are active from May until August with most records from July. The nest has been found to consist of four cells, with females being reared from the two innermost cells, and males from the two outer cells. A space, or vestibule, is present between the brood cells and the plug of the burrow entrance to the nest. Its prey consists of larvae of the chrysomelid beetle, *Zeugophora subspinosa*, F. and the gracillariid moth, *Caloptilia stigmatella*, F. The parasite, *Chrysis ignita* (L.) has been associated with this species in Europe (Spradbery, 1973).

Symmorphus crassicornis (Panzer, 1798) Plates 47 & 48

Black with yellow markings. Gastral tergites one to six usually with yellow apical markings. Length: female 10-15 mm, male 8-12 mm. Distributed in England from Cornwall to Kent and north to Worcestershire, Nottinghamshire and Norfolk, with an isolated record from Yorkshire and South Wales. Since 1970 it has undergone a significant decline. Distribution map in Edwards (1997) and Archer (2003a). Listed as RDB3 in Shirt (1987) and Falk (1991). Elsewhere, found in many parts of Europe and in Asia from Turkey to Manchuria. From the few records available adults are active from late June until early August and sometimes during early June and late August. Its prey consists of larvae of the chrysomelid beetle, *Chrysomela populi*, L. which is found on aspen (Guichard, 1972). The parasite, *Chrysis fulgida* L. has been associated with this species in mainland Europe (Schneider, 1991) and Britain.

Symmorphus gracilis (Brullé, 1832) Plates 49 & 50

Black with yellow markings. On female gastral tergites one to four or five usually with apical yellow markings, on male gastral tergites one to six usually with apical yellow markings. Length: female 8-12 mm, male 7-10 mm. Distributed from Cornwall to Kent and north to Yorkshire and Wales. Distribution map in Edwards (1997) and Archer (2003a). Elsewhere, found in many parts of Europe and east to central Asia. Males are active mainly during June and July, rarely May and August. Females are active mainly during June and July and sometimes at the end of May and rarely during August. Its prey consists of larvae of the chrysomelid beetle, *Chrysomela populi*, L. and a weevil, *Cionus hortulanus*, Geoffroy which is associated with figwort.

Genus **Ancistrocerus** Wesmael, 1836

This genus consists of about 21 species in Europe with nine British species.

Key to Species of *Ancistrocerus*

1. Posterior and lateral faces of the propodeum in part without surface structure, these parts smooth (Fig. 65) .... ***antilope*** (p. 38)

- Posterior and lateral faces of the propodeum with surface structure not smooth 2

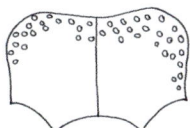

Figure 65. View of posterior face of propodeum of *Ancistrocerus antilpope*.

2. Surface of the second gastral sternite concave or flat posterior to the ridged furrow. Surface of the second gastral sternite does not abruptly slope into the ridged furrow. (View gaster ventrally to see these characters). Ridges in the furrow about 1.5 times longer than the distance between them (except for *A. quadratus* where about twice as long as the distance between them). The central ridges and most of the lateral ridges of similar length (less clear for *A. quadratus*) (Fig. 66) 3

- Surface of the second gastral sternite either convex posterior to the ridged furrow or the surface of the second gastral sternite abruptly slopes into the ridged furrow. (View gaster ventrally and laterally to see these characters). Ridges in the furrow at least twice as long as the distance between them. The longest ridges are the central ones (Fig. 67) 5

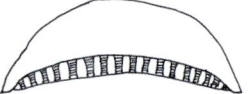

Figure 66. Ventral view of second gastral sternite of *Ancistrocerus parietum, A. gazella* and *A. quadratus.*

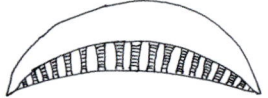

Figure 67. Ventral view of second gastral sternite of remaining species of *Ancistrocerus (A. nigricornis, A. oviventris, A. parietinus, A. scoticus* and *A. trifasciatus*).

3. Posterior face of the propodeum with many diagonal narrow ridges close to each other (Fig. 68). Ridges on the anterior face of the first gastral tergite with a deep central notch (Fig. 69). Yellow band on the first gastral tergite often gradually widened towards the sides (Fig. 70) .... ***parietum*** (p. 40)

- Posterior face of the propodeum without many diagonal narrow ridges (Fig. 71). If ridges are visible they are fewer in number and widely separated. Ridge on the anterior face of the first gastral tergite often without a central notch, but if a notch is present, it is shallow (Fig. 72). Yellow band on the first gastral tergite often abruptly expanded at the sides (Fig. 73) 4

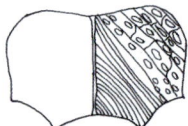

Figure 68. View of posterior face of propodeum of *Ancistrocerus parietum.*

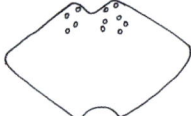

Figure 69. View of anterior face of first gastral tergite of *Ancistrocerus parietum.*

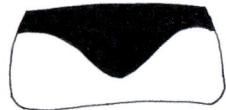

Figure 70. Dorsal view of first gastral tergite of *Ancistrocerus parietum.*

Figure 71. View of posterior face of propodeum of *Ancistrocerus gazella* and *A. quadratus.*

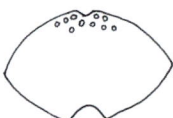

Figure 72. View of anterior face of first gastral tergite of *Ancistrocerus gazella* and *A. quadratus.*

Figure 73. Dorsal view of first gastral tergite of *Ancistrocerus gazella* and *A.quadratus.*

4. Antennal scape covered in large punctures, but with small punctures restricted to the apical part of the scape (Fig. 74). Surface of the second gastral sternite concave posterior to the ridged furrow. (View gaster ventrally to see this character). Central and lateral ridges (except extreme edge) in the furrow of similar length .... *gazella* (p. 38)

- Antennal scape covered entirely in large and small punctures (Fig. 75). Surface of the second gastral sternite flat posterior to the ridged furrow. (View gaster ventrally to see this character). Central ridges in the furrow longer than the lateral ridges ... *quadratus* (p. 40)

Figure 74. View of scape of *Ancistrocerus gazella*.   Figure 75. View of scape of *Ancistrocerus quadratus*.

5. Weakly convex surface of the second gastral sternum gradually merges into the ridged furrow. (View gaster ventrally to see this character) 6

- Surface of the second gastral sternum abruptly meets the ridged furrow, giving rise to a "cliff-like" structure. (View gaster ventrally and laterally to see this character) 7

6. Yellow band or spot on each of the anterior three or four gastral tergites. Gaster relatively more elongate, the first gastral tergite more than half as long as wide, when measured from the anterior ridge to the posterior margin of the tergite, and its greatest width .... *trifasciatus* (p. 41)

- Yellow band or spot on each of the anterior six gastral tergites. Gaster relatively less elongate, the first gastral tergite about half as long as wide, when measured from the anterior ridge to the posterior margin of the tergite, and its greatest width .... *parietinus* (p. 40)

7. Surface of the second gastral sternite behind the "cliff" concave. (View gaster ventrally to see this character). (Yellow band or spot on each of the anterior five or six gastral tergites). (The anterior margin of the clypeus of the male with a half-circular notch) .... *nigricornis* (p. 39)

- Surface of the second gastral sternite behind the "cliff" weakly convex or flat. (View the gaster ventrally to see this character). 8

8. Yellow band or spot on each of the anterior five or six gastral tergites. The anterior margin of the clypeus of the male with a notch deeper than a half-circle (Fig. 76) ... *oviventris* (p. 39)

- Yellow band or spot on each of the anterior three or four gastral tergites. The anterior margin of the clypeus of the male less deep, like a half-circle (Fig. 77) .... *scoticus* (p. 40)

Figure 76. View of male clypeus of *Ancistrocerus oviventris*.   Figure 77. View of male clypeus of *Ancistrocerus scoticus*.

Species accounts

Generally tube-dwellers, except *A. oviventris* and *A. scoticus*, in plant stems, *e.g.* bramble and elder, and a variety of other cavities. Each nest consists of two to eight linearly-arranged cells separated by clay partitions. After an egg is laid, several paralysed larvae, usually lepidopteran, are forced into the cell. The egg hatches after a few days and the larva consumes the prey in one to two weeks. When fully fed the larva rests for a day or two before it voids waste products from the gut. The larva spins a cocoon and enters winter diapause as a prepupa. When winter diapause is absent the prepupal stage lasts for about one week, followed by pupation. The adult emerges after about two weeks later. Adults have been recorded visiting flowers of sea-holly, bramble, raspberry, goldenrod, Michaelmas daisy, dwarf elder, hogweed, privet, nightshades and thistles.

Ancistrocerus antilope (Panzer, 1798) Plates 16 & 17

Black with yellow markings. Gastral tergites one to six usually with apical yellow markings. Length: female 13-16 mm, male 10-13 mm. Distributed from Cornwall to Kent and north to Cumbria and Northumberland, Wales and the Channel Islands (Guernsey). There is also a record from Scotland, 'near Hillieslands, Roxburghshire' during 1882 (Clark, 1924), but the grid reference for this locality is unknown. Since 1970, its range has apparently decreased to seven 10 km squares. Map of British distribution in Edwards & Telfer (2001) and Archer (2003a). Listed as RDB3 in Shirt (1987) and Falk (1991). Overseas, found in many parts of Europe, North Africa, Asia from Turkey to Japan and North America. Associated with a variety of habitats where nesting sites and prey are available. Adults are active mainly during June and July, less frequently during May and rarely during August and September. Besides nesting in plant stems, nests found in wood, old mortar walls and in the cavity of a window of a commuter train, even though the train made several journeys each day. Its prey usually consists of lepidopteran larvae from several families, e.g. Pyralidae and Gelechiidae, beetle larvae of Chrysomelidae and more rarely sawfly larvae. Parasitized by *Chrysis pseudobrevitarsis* Linsenmaier and *C. longula* Abeille de Perrin (Morgan, 1984). The mite *Kennethiella trisetosa* (Cooreman) feeds on the pupal stage without causing death. On emergence of the adult wasp the mites, which are mainly on the males, cling to the propodeum before moving into the genital chamber. During copulation of the wasps the mites are passed to the female wasps (*in* Spradbery, 1973).

Ancistrocerus gazella (Panzer, 1798) Plates 18 & 19

Yarrow (1954) introduced this species to the British list giving characteristics by which it can be separated from *A. parietum*. Black with yellow markings. Female with gastral tergites one to four or five usually with apical yellow markings, male with gastral tergites one to four or six usually with apical yellow markings. Length: female 8-12 mm, male 7-9 mm. Distributed in Ireland, Wales and England from Cornwall to Kent, north to Stockton-on-Tees. They are two isolated records from Scotland (Edinburgh, Isle of Mull). Also recorded from Lundy Island, Isles of Scilly and the Channel Islands. Distribution map in Edwards (1998) and Archer (2003a). Elsewhere, found in many parts of Europe, Asia east to Iran and also North Africa. Found in a wide variety of habitats including river banks, coastal areas and open urban, parkland and wooded areas. Adults are active usually from June until August, and sometimes during September (particularly the females), and rarely during May and October. Parasitized by *Chrysis* species.

Ancistrocerus nigricornis (Curtis, 1826) Plates 20 & 21

Previously known as *A. callosus* (Thomson, 1870). Black with yellow markings. Female with gastral tergites one to five usually with apical yellow markings, male with gastral tergites one to six usually with apical yellow markings. Length: female 9-13 mm, male 6-10 mm. Distributed from Cornwall to Kent and north to Cumbria and Northumberland (Gateshead) and Scotland (Roxburghshire and Aberdeenshire). Richards (1980) also gives Raasay in Scotland but no further details have been found. Also recorded from the Isle of Man and the Channel Islands. There is a single record from Ireland (Wicklow). Since 1970, it has undergone a significant decline, particularly in the north of England. Distribution map in Edwards & Telfer (2001) and Archer (2003a). Elsewhere, found in many parts of Europe, North Africa, and Asia from Turkey to Japan. Found in a wide range of habitats including gardens. Males active from June until September but mainly during July and August. Females active from February until October with a small peak of abundance during May and a larger peak of abundance during August. Males and females emerge in late summer, when mating occurs; the males then die and the females overwinter as adults, to reappear in the spring. Besides nesting in plant stems, nests are found in wood (dead logs and tree trunk, tree-stumps, telegraph poles and fences) and old mortar. Prey consists of small brown and green microlepidopteran larvae, e.g. Tortricidae. In mainland Europe parasitized by *Chrysis schencki* Linsenmaier (Schneider & Leclercq, 1987) and *C. ignita* (L.) (Kunz, 1994).

Ancistrocerus oviventris (Wesmael, 1826) Plates 22 & 23

Previously known as *A. pictus* (Curtis, 1826). Black with yellow markings becoming whitish yellow in Scottish and Irish specimens. Tibiae and tarsi of female reddish colouration. Female with gastral tergites one to six usually with apical yellow markings, male with gastral tergites one to five usually with apical yellow markings. Length: female 11-14 mm, male 6-12 mm. Distributed throughout England, Wales, Scotland and Ireland including the Isle of Man, Outer Hebrides and the Channel Islands. Since 1970, this species has undergone a significant decline, particularly in central and eastern England. Distribution map in Edwards (1998) and Archer (2003a). Elsewhere, found in many parts of Europe, North Africa, and Asia from Turkey to Japan. Found in a wide range of habitats including moorland, lowland heathland, clay and sandy woodlands, parklands, limestone and chalk quarries, calcareous grassland, coastal cliffs, sand dunes and urban areas (gardens and cemeteries). Adults mainly active from May until July, sometimes during August, and rarely in April, September and October. This species is a mud-dauber building its clay nests on walls, stone columns, concrete blocks, and rocks; often using crevices and holes as starting points for nest building. It has also been found nesting in a door lock (Sheppard, 1926) and in the holes of an oak cribbage board (Blair, 1943). Initially a row of cells is built on the substrate with additional cells placed on top. The nest usually consists of three to five cells (up to 14 cells recorded) which are covered by a layer of mud as a protective camouflaging layer. Usually four or five prey items are placed in each cell. Its preys consist of microlepidopteran larvae, usually Tortricidae. Parasitized by *Chrysis ruddii* Shuckard (Morgan, 1984) and probably *C. ignita* (L.) or *C. impressa* Schenck (Bignell, 1882). The braconid, *Meteorus pallipes* (Wesmael), and the tachinid fly, *Phytomyptera cingulata* (Robineau-Desvoidy) (as *Craspedothrix zonella* Zetterstedt) have been reported as parasites of the eumenid's prey (Blair, 1943).

Ancistrocerus parietinus (Linnaeus, 1758) Plates 24 & 25

Black with yellow markings. Gastral tergites one to six usually with apical yellow markings. Length: female 12-15 mm, male 9-12 mm. Distributed throughout England and Wales, including the Isle of Man, and into Scotland north to Aberlour (Banffshire). Widely recorded in Ireland and there is one old record from the Channel Islands. Distribution map in Edwards (1998) and Archer (2003a). Elsewhere, found in many parts of Europe and Asia east to Japan. Found in a wide range of habitats including gardens. Adults active usually during June and July, sometimes during May and August, and rarely during March, April and September. Besides nesting in plant stems, nests have been found in the central hole of a cotton reel, gap between books and the back of a bookcase and a hole in a wooden clothes line prop. Prey consists of microlepidopteran larvae and beetle larvae of Chrysomelidae. Parasitized by the chrysid wasps, *Chrysis impressa* Schenck and *C. longula* Abeille de Perrin (Morgan, 1984).

Ancistrocerus parietum (Linnaeus, 1758) Plates 26 & 27

Black with yellow markings. Gastral tergites one to six usually with apical yellow markings. Length: female 8-12 mm, male 7-11 mm. Distributed throughout England and Wales including the Isle of Man, Lundy Islands, the Channel Islands and into Scotland north to Aviemore (Moray) and Aberdeen. Also recorded from Ireland. Distribution map in Edwards (1998) and Archer (2003a). Elsewhere, found in many parts of Europe, North Africa and Asia from Turkey to Manchuria and introduced into North America. Found in a wide range of habitats including river banks, coastal areas and open urban, parkland and wooded areas. Adults active usually from June until August, sometimes during May and September, and rarely during October. Besides nesting in plant stems, nests found in straw of a thatched roof, disused burrows of wood-boring insects, a disused embryo nest of the social wasp, *Vespula vulgaris* (L.), and the disused nest of a mason wasp, *Odynerus* species. Parasitized by the *Chrysis ignita* (L.) (Morgan, 1984).

Ancistrocerus quadratus (Panzer, 1799)

Previously known as *A. claripennis* (Thomson, 1874). Black with yellow markings. Female with gastral tergites one to five usually with apical yellow markings, male with gastral tergites one to four or six usually with apical yellow markings. Length: female 10-12 mm, male 7-11 mm. Recorded from four 10 km squares in Devon (1958-1969), one 10 km square in Herefordshire (pre-1859), two 10 km squares in Leicestershire and one 10 km square in Nottinghamshire (1905-1944). Probably now extinct. Listed as RDB3 in Shirt (1987) and Falk (1991). Distribution map in Edwards (1998) and Archer (2003a). Elsewhere, found in many parts of Europe, rarely southern Europe, and parts of Asia (Israel, Turkey). One male was recorded during June and two females, one during August and one during September. Habitat unknown but probably found in a wide range of habitats. Besides nesting in plant stems, nests found in an old razor case, folds in a piece of paper, bore of a flute and on a shelf behind some books.

Ancistrocerus scoticus (Curtis, 1826) Plates 28 & 29

Previously known as *A. albotricinctus* (Zetterstedt, 1838) and *A. trimarginatus* misidentified. Black with yellow markings becoming whitish yellow in Scottish and Irish specimens. Tibiae and tarsi of female reddish colouration. Female with gastral tergites one to three usually with apical yellow markings, male with gastral tergites one to four usually with apical yellow markings. Length: female 10-12 mm, male 7-9 mm. Distributed throughout England, Wales and Scotland including the Isle of Man, the Isles of Scilly and coastal Ireland. It is now mainly associated with the coast with a significant decline from inland areas. Distribution map in Edwards (1998) and Archer (2003a). Elsewhere, found in many parts of Europe, North Africa and Asia from Turkey to Korea and Kamchatka. Found in a wide range of habitats including moorland, lowland heathlands, clay and sandy woodlands, parkland, limestone and sandstone quarries, limestone grassland, urban areas (gardens), coastal cliff and heaths, marshes, sand dunes and coastal shingle. Adults are usually active from June until August, sometimes during May and September, and rarely during April. Nests of clay cells are found in crevices as in rocks and bridges, in holes in pebbles and mine slag, among the bark of pine trees, in dead stems of elder and reeds. Can also burrow in flat bare areas making its cells flush with the surface of the ground (Julliard, 1950). Parasitized by *Chrysis ignita* (L.) and probably *C. rutiliventris* Abeille de Perrin (Morgan, 1984).

Ancistrocerus trifasciatus (Müller, 1776) Plates 30 & 31

Previously known as *A. trimarginatus* (Zetterstedt, 1838). Black with yellow markings. Female with gastral tergites one to three or four usually with apical yellow markings, male with gastral tergites one to four usually with apical yellow markings. Length: female 10-12 mm, male 7-10 mm. Distributed throughout England and Wales, including the Isle of Man, the Channel Islands and southern Scotland. Also found in Ireland. Distribution map in Edwards (1998) and Archer (2003a). Elsewhere, found in many parts of Europe and Asia from Turkey to Japan. Found in a wide range of habitats and often associated with marshy places. Adults usually active from June to August, sometimes during May and September, and rarely during April. Besides nesting in plant stems, also found nesting in disused galls of *Andricus kollari* (Hartig). Jørgensen (1942) gives further information about nesting characteristics. Prey consists of small lepidopteran larvae and sometimes beetle larvae of Chrysomelidae. Parasitized by *Chrysis angustula* Schenck, *C. impressa* Schenck and *C. mediata* Linsenmaier (Morgan, 1984).

Genus **Microdynerus** Thomson, 1874

This genus consists of about 20 species in Europe with one British species.

Species account

Microdynerus exilis (Herrich-Schäffer, 1839) Plate 38

Black with whitish yellow markings. Gastral tergites one and two only with apical whitish-yellow markings. Length: female 6-8 mm, male 5-6 mm. Introduced to the British list by Jones (1937) from Hampshire. By 1958 it had spread to Wiltshire, Isle of Wight and West Sussex (Richards, 1958). By 1973 it had spread to Dorset, Buckinghamshire, Middlesex and Surrey (Else, 1973). By 1991 it had spread to East Sussex, East and West Kent, Oxfordshire and West Norfolk (Falk, 1991). Currently distributed from Dorset to Kent and north to Herefordshire, Warwickshire and West Norfolk. Regarded as a Scarce (Nb) species by Falk (1991). Distribution map in Edwards & Telfer (2001) and Archer (2003a). Elsewhere, found in many parts of Europe. Found in a wide range of habitats including woodland, parkland, gardens, heathland edge, chalk downland, gravel pits and coastal sites. Adults are usually active during June and July, sometimes during August and rarely during September. It is a tube-dweller nesting in small beetle holes in dead wood, including old fence posts, and sometimes in bramble stems. Danks (1971) described a nest consisting of six linear cells with an outer plug of fine pith and fairly coarse sand grains. The outer four cells produced males and the inner two cells produced females. A female has been observed carrying a mud pellet in the mandibles, probably used to form the partitions between the cells. Its prey consists of weevil (Curculionidae) larvae. Adults have been found on flowers of hawk's-beard, hogweed and mayweed. Overseas, Kunz (1994) gives *Chrysis gracillima* Förster as a parasite and Wilson (pers. comm., 2013) associated *M. exilis* with *C. gracillima* in an urban park at Colchester.

Genus **Pseudepipona** de Saussure, 1855

This genus consists of eight species in Europe with one British species.

Species account

Pseudepipona herrichii (de Saussure, 1856) Plates 43 & 44

Previously known as *P. variegata* misidentified and *P. basalis* (Smith, 1857). Black with whitish yellow markings. Gastral tergite one with lateral large reddish markings. Legs reddish coloured. Female with gastral tergites one to five usually with apical yellow markings, male with gastral tergites one to six usually with apical yellow markings. Length: female 11-13 mm, male 9-11 mm. This species is restricted in Britain to a few sites in the Poole Basin of south-east Dorset. This species regarded as RDB2 by Shirt (1987) and RDB1 by Falk (1991). Distribution map in Edwards & Telfer (2001) and Archer (2003a). Elsewhere, found in many parts of Europe, North Africa, Asia east to China and far eastern Russia and North America. Restricted to lowland heathlands where it requires resources of soil with a clay content exposed to the sun (as a nesting site), open water (to assist in nest building) and heathland rich in bell heather, *Erica cineraria* L., at the early to middle heathland stage (for both nectar and prey foraging). The nesting site may be on a vertical or flat surface. The water is needed to soften the clay when digging the nest. It will also take nectar from the flowers of thyme. As both sexes have short tongues, the flowers are invariably raided by biting through the base of the corolla. Adults usually active from mid-June until early August, occasionally to mid-August. The underground nests are in dense aggregations being excavated in bare ground in sunny situations. The burrows are shallow (only a few centimetres deep), and may contain up to three cells (Spooner, 1934). The cells are stocked with eight to 20 tortricid moth larvae, *Acleris hyemana* (Haworth). The completed nest burrow is sealed with a plug of moistened clay. Further information is given in Morrison, 1991; Edwards & Roberts, 1995; Roberts & Else, 1997, Roberts, 1998, 1999, 2000, 2001; Neal, 2006, Dieck & Neal, 2007.

Genus **Euodynerus** Dalla Torre, 1904

This genus consists of about twelve species in Europe with one British species.

Species account

Euodynerus quadrifasciatus (Fabricius, 1793) Plates 34, 35 & 60

Previously known as *E. tomentosus* (Thomson, 1870). Black with yellow markings. Female with gastral tergites one to four usually with apical yellow markings, Length: female 11-12 mm, male 9-11 mm. Found in Devon, Dorset, Berkshire and Surrey with a nineteenth century record from West Gloucestershire. Richards (1958) and Else (1992) give details of early records. This species regarded as RDB3 by Shirt (1987) and RDB2 by Falk (1991). Distribution map in Edwards & Telfer (2001) and Archer (2003a). Elsewhere, found in many parts of Europe, North Africa and Asia east to Japan. Found in the habitats of limestone quarries, pebble beaches and slopes of coastal cliffs. Adults are usually active during June and July, sometimes during May and August. It is a tube-dweller nesting in holes in beach pebbles and in large stones, rocks and probably in burrows in dead wood (Else, 2006; Else & Roberts, 2006). Prey consists of microlepidopteran (Tortricidae) and beetle (Chrysomelidae) larvae. Adults visit flowers of dewberry and wall cotoneaster. Kunz (1994) gives *Chrysis longula* Abeille de Perrin as a parasite.

Genus **Pterocheilus** Klug, 1805

This genus consists of one species, *Pterocheilus phaleratus*, which is a subterranean nester digging a short vertical burrow in sandy soil which ends in a horizontally placed cell. After the prey consisting of caterpillars of the moth family Psychidae have been placed in the cell, the burrow is back-filled with sand (Nielsen, 1942).

Species account

One Channel Islands species.

Pterocheilus phaleratus (Panzer, 1797)

Black with yellow markings. Female with gastral tergites one to five usually with apical yellow markings, male with gastral tergites one to six usually with apical yellow markings. Length: female 7-8 mm, male 7-8 mm. There is one record of a male from Jersey during April and possibly a vagrant. Overseas, found in many parts of Europe and North Africa.

Genus **Gymnomerus** Blüthgen, 1938

This genus consists of one species in Britain with several subspecies in Europe.

Species account

Gymnomerus laevipes (Shuckard, 1837) Plates 36 & 37

Black with yellow markings. Female with gastral tergites one to four usually with apical yellow markings, male with gastral tergites one to five usually with apical yellow markings. Length: female 8-11 mm, male 9-10 mm. Distributed in England from Cornwall to Kent and north to Glamorgan, Herefordshire, Worcestershire, south Nottinghamshire, south Lincolnshire, East Norfolk and South Wales. Distribution in Edwards & Telfer (2001) and Archer (2003a). Elsewhere, found in many parts of Europe, North Africa and Asia from Turkey to far eastern Russia. Found in a wide range of open habitats including open woodland. Adults are active from May until September, with males mostly found from late May and June and females during June and July. A tube-dweller forming a series of 4-12 linearly arranged cells separated by clay partitions in hollow plant stems such as bramble, burdock, elder and thistles. The cell cavity has been found to be lined by a compact cement of fine sand or chalk (Chambers, 1944, 1949). Each cell is provided with about 20 weevil larvae (*Hypera*). The adults visit the flowers of wild angelica, ash, bramble, buttercup, carrot, figwort and spurge. The chalcid parasitoid, *Melittobia acasta*

(Walker) and the ichneumonid parasitoid, *Bathyplectes* (=*Canidiella*) species are associated with *Gymnomerus*. The female *Melittobia* enters a cell of *G. laevipes*, punctures the cuticle of the larva with its ovipositor and feeds on the haemolymph released. Eggs are laid on the cuticle of the larval host. On hatching, the young wasp larvae pierce the cuticle of their host with their mandibles and also feed on the haemolymph. The female *Bathyplectes* is not a parasite of *G. laevipes*, but is an endoparasite of its prey *Hypera* (*in* Spradbery, 1973).

Genus **Odynerus** Latreille, 1802

This genus consists of 20 species found in Europe and with four British species.

The following life-history description is based on that of *O. spinipes*, but it is probably similar for other species. Mating takes place shortly after the adults emerge, followed by the females searching for nesting sites. The digging spot is wetted with water and a cluster of cells is excavated. The excavated material is first used to build a 'chimney'. The function of the chimney is unknown but it could prevent the entry of rain into the burrow which is situated in rather exposed situations. It has also been suggested that the chimney deters potential cleptoparasites and parasitoids. However, Uffen (1998), observed workers of *Lasius niger* (L.) remove the weevil prey from a nest of *O. spinipes*. The ants exited between the holes of the sand chimney. The female hunts for prey which when found is immobilised by stinging and malaxation. The egg hatches in a few days and resulting larva eats the prey provided in a matter of weeks. The overwintering stage is probably the prepupa.

Key to the species of *Odynerus*

1. Female with posterior surface of the metanotum covered with small punctures (Fig. 78); metanotum black. Male with the middle femur having two deep notches and three projections; no spine present on the gena and middle coxa 2

- Female with posterior surface of the metanotum without punctures (Fig. 79); metanotum with an anterior yellow band. Male with the middle femur without deep notches and projections; spine present on the gena (Fig. 80) and middle coxa 3

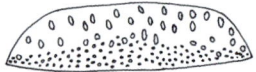

Figure 78. View of posterior surface of metanotum of *Odynerus spinipes* and *O. melanocephalus*.

Figure 79. View of posterior surface of metanotum of *Odynerus reniformis* and *O. simillimus*.

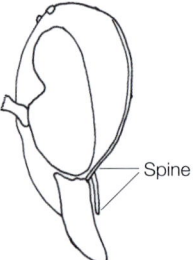

Figure 80. Lateral view of head of *Odynerus reniformis* and *O. simillimus*.

2. Female with the head, thorax and propodeum covered with black hair. Male with the central projection on the middle femur truncate (Fig. 81) .... *spinipes* (p. 46)

- Female with the head, thorax and propodeum covered with pale hair. Male with the central projection on the middle femur bluntly pointed (Fig. 82) .... *melanocephalus* (below)

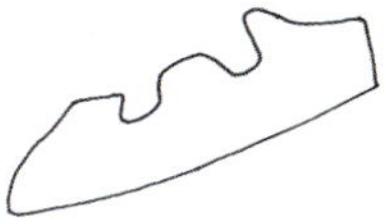

Figure 81. Lateral view of male middle femur of
Odynerus spinipes.

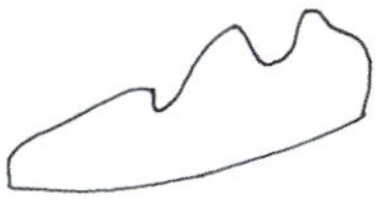

Figure 82. Lateral view of male middle femur of
Odynerus melanocephalus.

3. Female with the central fine punctures on the second gastral sternite close together so that the distance between the punctures is more-or-less equal to the diameter of the punctures. Male with the middle coxal spine parallel-sided so that it appears longer Fig. 83) ....
.... *reniformis* (p. 46)

- Female with the central fine punctures on the second gastral sternite widely separated so that the distance between the punctures is greater than the diameter of the punctures. Male with the middle coxal spine not parallel-sided so that it appears shorter (Fig. 84) ....
.... *simillimus* (p. 46)

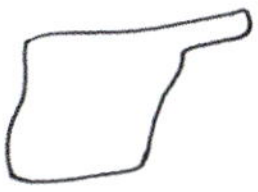

Figure 83. View of male middle coxal spine of
Odynerus reniformis.

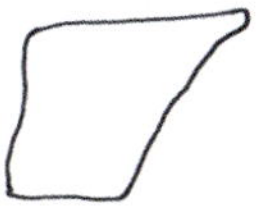

Figure 84. View of male middle coxal spine of
Odynerus simillimus.

Species accounts

Odynerus melanocephalus (Gmelin *in* Linnaeus, 1790) Plates 39 & 40

Black with whitish-yellow markings. Female with gastral tergites one to four usually with apical whitish-yellow markings, male with gastral tergites one to six usually with apical whitish-yellow markings. Length: female 8-10 mm, male 10 mm. Distributed in England from Devon to Kent and north to Northamptonshire and South Wales. Since 1970, it has undergone a significant decline. Falk (1991) listed this species as nationally Scarce (Na). Distribution map in Edwards (1997) and Archer (2003a). Elsewhere, found in many parts of Europe, North Africa and Asia (Turkey, Israel, Iran). Found in a wide range of open habitats including grasslands, woodland, soft rock cliffs, landslips, saltmarsh margins and post-industrial sites (brick and sand pits and railway embankments). Females are usually active during June and July and rarely during April, May and August. Adult males are usually active during May and June, sometimes during July and rarely during August. Nests are dug in bare or sparsely vegetated areas in soils with high clay content. The chimney associated with the burrow entrance is rather short, about 1 cm long. Its prey consists of weevil larvae, *Hypera postica* (Gyllenhall) found on black medick. The prey is also found on common bird's-foot-trefoil (Edwards, 2007; Hunnisett, 2006; Wright, 2007). Adult visits the flowers of black medick, wild carrot, clover and speedwell. Information was obtained from Falk (pers. comm., 2005) and Falk (2005) which gives further details. Parasitized by *Spinola neglecta* (Shuckard) (Morgan, 1984; Falk, pers. comm., 2005).

Odynerus spinipes (Linnaeus, 1758) Plates 41 & 42

Black with yellow markings. Female with gastral tergites one to four usually with apical yellow markings, male with gastral tergites one to four or five usually with apical yellow markings. Length: female 10-12 mm, male 10-12 mm. Distributed throughout England and Wales and into southern Scotland with an isolated record from northern Scotland. Also recorded from the Isle of Man, Channel Islands and there are two old records from Ireland. Since 1970 it has undergone a significant decline. Distribution map in Edwards (1997) and Archer (2003a). Elsewhere, found in northern and middle Europe, North Africa and Asia from Turkey to far eastern Russia. Found in a wide range of open habitats digging nests in vertical, usually clay surfaces, but also sandy surfaces, and even in the mortar of old walls (Falk, pers. comm., 2005). Females in northern England are active usually during May and June, rare during August. In southern England females are active mainly during June and July and sometimes during May and August. In northern and southern England males are active mainly during June but also during May and July. During mating the male places himself on the back of the female securing her wings in the spaces between the deep notches of the middle femur and held in this position by the tibia (Chapman, 1870). The nest consists of a cluster of five to six cells. The chimney associated with the burrow entrance is rather long, about 3 cm, which curves over and downwards. Its prey consists of weevil larvae (*Hypera*) with up to about 30 larvae being placed in each cell. The prey has been found on common bird's-foot-trefoil (Edwards, 2007). Adults usually visit flowers with a short corolla and accessible nectaries. Extrafloral nectaries and honeydew from aphids may also be taken. The parasitic wasp, *Chrysis viridula* L., acts as a parasitoid feeding on the mature larva or pupa of its host (Chapman, 1877). *Pseudospinola neglecta* (Shuckard) is also a parasite (Chapman, 1877; Morgan, 1984).

Odynerus reniformis (Gmelin *in* Linnaeus, 1790)

Black with pale yellow markings. Female with gastral tergites one to five usually with apical yellow markings, male with gastral tergites one to six usually with apical pale yellow markings. Length: female 12-13 mm, male 9-12 mm. Recorded from Hampshire and Surrey; the last was during 1909 from the New Forest. Also recorded from the Channel Islands. Both Shirt (1987) and Falk (1991) regard this species extinct (RDB1+) in mainland Britain. Distribution map in Edwards (1997) and Archer (2003a). Elsewhere, found in many parts of Europe, North Africa and Asia from Turkey to far eastern Russia. Mainly found on heathlands with bare soil, but also railway embankments. From the few records available adult females are active from May until August and males from May until July. Nesting sites are on level ground and vertical surfaces. It has also been found nesting in the walls of a ruined cottage (Morice, 1906). The chimney associated with the burrow entrance is about 2 cm long and curved when on level ground and slightly curved downwards when projecting from a vertical surface. The nest consists of one to three cells. Its prey consists of lepidopteran and sawfly larvae and, in one recording, a weevil larva (*Hypera*) (Nielsen, 1932). Parasitized by *Spinola neglecta* (Shuckard) (Morgan, 1984) and *Chrysis viridula* L. has been associated with this species in mainland Europe (Spradbery, 1973; Kunz, 1994).

Odynerus simillimus Morawitz, 1867

Black with whitish-yellow markings. Female with gastral tergites one to five usually with apical whitish-yellow markings, male with gastral tergites one to six usually with apical whitish-yellow markings. Length: female 11-12 mm, male 11 mm. Distributed in the coastal areas of East Anglia (Essex, Suffolk and Norfolk). Until its rediscovery in 1986

(Archer, 1989b) it was considered extinct. It was originally first recorded during 1902 (Saunders, 1903). Shirt (1987) regarded it as extinct (RDB1+) but Falk (1991) considered it a RDB1species. Distribution map in Edwards (1997) and Archer (2003a). Elsewhere, found in middle and eastern Europe and Asia from Turkey to far-eastern Russia. It is associated with fenland and coastal wetlands. Found nesting near coastal areas in friable clayey, silty or sandy/clay soils on bare ground or at bases of open tussocky grass. A near-by water source is needed to moisten the clay. In particular, nests have been found in near vertical and sloping banks, arable field margins, edge of footpaths, dumped ditch dredging and uprooted tree root bole (Harvey, 2001, 2002; Scott, 2003). Nest aggregations of five to 13 nests have been found (Harvey & Scott, 2001). From the few available records males are active during June and July and females from June until August. Its prey consists of *Hypera* weevil larvae which feed on wetland umbellifers found in fen and ditches. One nest was found to consist of four cells. The chimney associated with the burrow entrance is only about 0.5 cm long (Blüthgen, 1961; Spradbery, 1973).

Subfamily Polistinae

Worldwide in distribution with about 950 species in 26 genera. Nine species in Europe with two species recorded in Britain.

Genus **Polistes** Latreille 1802

Polistes has a worldwide distribution consisting of about 203 species (Carpenter, 1996). The adults vary in length from 10-35 mm. They are largely black or dark brown, yellow and black or rust coloured. The first gastral segment is short and broad, appearing more or less triangular from above.

The nest of most *Polistes* species is a single exposed comb usually supported by a single pedicel. The comb may be vertical or horizontal in orientation. A few species build multiple combs and multiple pedicels. The nest material consists of long woody fibres from weathered wood and other sources including paper from old nests and plant hairs.

Nests are usually built in protected aerial sites but are clearly visible. Such sites include tree trunks and the undersides of leaves, and man-made structures, such as the undersides of eaves and roofs of buildings and under bridges. A few nests are concealed in holes in trees or cavities covered by grass on the soil surface or inside buildings.

The colony cycle may be divided into four phases: founding, worker, reproductive and intermediate (Reeve, 1991). Nest founding is by a single foundress or jointly with several foundresses. One of these foundresses may or may not become dominant and lay most of the eggs. The other auxiliary or subordinate foundresses may leave and join another colony, attempt to usurp another colony or adopt a newly abandoned colony. If the subordinate foundresses remain, they may share the egg laying, supersede the dominant foundress or become non-egg laying helpers. For species from temperate climates this phase starts from early to late spring while for species from tropical climates this phase can start at any time of the year. Temperate species show synchronised life cycles and tropical species unsynchronised life cycles.

At the start of the worker phase, when the first workers emerge in many species, the foundress associations break up with the subordinate foundresses disappearing. Males may appear with the first workers. A male may mate with a worker which then supersedes the queen or establishes a satellite nest near the original nest. For species from temperate climates this phase starts from late spring until early summer. The reproductive phase begins with the emergence of the males and future foundresses or gynes. Caste differences between workers and gynes are small, being mainly behavioural. For species from temperate climates this phase starts from early to mid-summer.

The intermediate phase starts with the decline of the colony which, for species from temperate climates, starts from early to mid-autumn. Colony decline is often associated with the disappearance of the queen or cessation of egg-laying. The reproductives disperse and mate. The males usually die fairly quickly while gynes (new queens) from temperate regions enter winter diapause before entering the founding phase. Gynes from tropical regions can directly enter the founding phase.

Most nests have fewer than 500 cells at maturity and rear 200 or fewer adults. Exceptionally large nests exceed 1500 cells.

In Europe, three species are social parasites, with no worker caste. A parasite female invades a host's nest usually in late June when only a few workers are present. The parasite female becomes the dominant female and sole egg-layer while the host queen either disappears or becomes a worker. The parasite queen does not kill the host queen but rather subordinates her to a non-laying helper.

Diagnosis of British species

All species are social or social parasites.

Mandibles short, overlapping transversely when fully closed.

Hind wing with a short jugal lobe.

Middle tibia with two spurs.

Claws simple.

First gastral segment narrowed anteriorly with the anterior and dorsal surfaces continuous.

Identification of the European species found in Britain

Separation of the two species, particularly the females, is not easy and reference should first be made to identified specimens and Dvořák & Roberts (2006).

Females of *P. dominula* typically have black coloured mandibles, a yellow mark on the lower part of the gena and the last gastral sternum partly or entirely yellow. Unfortunately, in light coloured forms the mandible has a yellow spot. Because this species has a more robust appearance the clypeus appears broader than long (but this is not supported by measurements).

Females of *P. gallicus* typically have a yellow mark on each mandible, the lower gena black and the last gastral sternum entirely black. Unfortunately, in light coloured forms the lower gena has a yellow mark and the last gastral sternum may only be partly black. Because this species has a less robust appearance the clypeus appears to be as wide as long (but this is not supported by measurements).

The male of *P. dominula* has the anterior margin of the clypeus dark with a central dark projection. In dorsal aspect the head behind the eyes is bulbous being convex.

The male of *P. gallicus* has the anterior margin of the clypeus translucent without a central dark projection. In dorsal aspect the head behind the eyes is not bulbous but straight, or slightly concave, and converging posteriorly.

Other species of *Polistes* from the Americas have been reported from Liverpool (Smith, 1875-6) found in wool warehouses, from Penzance (Smith, 1878) probably on ships from Brazil and from Glasgow (Cooter, 1976) found inside empty whisky barrels from the U.S.A. Richards (1978) may be used to attempt identification of American *Polistes*.

Species accounts

Polistes dominula (Christ, 1791)

Formerly known as *P. gallicus* and *P. dominulus*. Yellow and black as indicated in the identification text key. The first three segments of the antenna with dorsal dark markings with the rest of the flagellum yellow. Clypeus yellow without or with a variable size black marking. Length: queens 15-17 mm, workers 10-15 mm, males 10-17 mm. Richards (1980) gives records from Chandler's Ford, Hampshire (SU4320) during 1911 and from Wolsingham, Co Durham (NZ0737) during 1915 although the identity of the specimens is not known for certain. Richards (1980) gives a further two records when the identification of the specimens is certain: Mill Hill, Middlesex (TQ2192) during 1937 and Hither Green, Kent (TQ3874) during 1975. The Mill Hill specimen was found in lettuce and the Hither Green specimen in endive from Spain, indicating that these records are of vagrants imported with vegetables. Further records for vagrants are: Maidstone, Kent (TQ7655), January 1958, associated with a box of peaches (Clemons, 1993); Edinburgh (NT2573), December 1979, associated with cos lettuce (McKinley, 1980); Cheam, Surrey 1951 (TQ2463), in house, (Baldock, 2003); Harrogate, North Yorkshire (SE3055), April 2004, associated with Spanish vegetables (A. Snelson, pers. comm.). Recently, records probably indicate the occurrence of nests: Burham, north Kent (SU9383), several individuals seen August 1990 visiting flowers, also 1st September (Albertini, M., pers. comm.); Erith marshes, north Kent (TQ4979), September 1992, a male at flowers of fennel (Clemons, 1993); Thames Barrier, Essex (TQ4180), September 2006, several males flying and landing on the tops of twigs, a nest found in a hedge (Jones, R.A., pers. comm.); Thurrock, Essex (TQ5958), June 2006, a single female (Knowles, 2006); Rothley, Leicester (SK5812), 2007, a nest of *Polistes* of unknown species from a garage ceiling (BENHS Annual Exhibition 2007). A nesting site has been found at Ham House, Surrey (TQ1772) which has persisted for several years: 2003, about 20 males and females at flowers of fennel (Baldock, 2003); adult found in July 2004 indicating overwintering (Baldock, 2004); 2005 during mid-July, females observed taking wood from an old oak bench, nests found in the roof of an orangery, about six nests were seen from the top of a ladder but others were present (Baldock, 2005). The wasps were still present in numbers during 2006 and 20 or more workers seen hunting on 10th June by D. Baldock and on 13th August by A.J. Halstead. The site was not visited during 2007 (Baldock, pers. comm.). Except for the Leicester record, all the nesting sites are on the Thames estuary or Thames valley.

Overseas, found in many parts of Europe, North Africa and part of Asia although the status of Asian species awaits further study. Introduced into the U.S.A.

Polistes gallicus (Linnaeus, 1767)

Formerly known as *P. foederatus*, (Kohl) (or *P. foederata*) and *P. omissa*, (Weyrauch). Yellow and black as indicated in the identification text key. The first three segments of the antenna with dorsal dark markings while the rest of the flagellum yellow or may be slightly darkened. Clypeus yellow without or with a variable size black marking. Length: queens 14-15 mm, workers 10-13 mm, males 10-11 mm. This is a vagrant species with a male recorded in Bangor, Co. Down during September 1962 (McClenaghan, 1979).

Elsewhere, found in many parts of Europe, North Africa and part of Asia although the status of the Asian species awaits further study.

Subfamily Vespinae

Distributed throughout the Holarctic and Oriental regions with about 70 species in four genera (Archer, 1989a; Carpenter & Kojima, 1997). Three British genera with nine species.

The vespine social wasps have an annual life-cycle in the British Isles. Adults feed on material containing carbohydrates such as tree sap, nectar, honeydew and steal honey from colonies of bumblebees. Overwintering fertilised queens emerge in the spring and build a nest from wood fibres which are macerated and mixed with saliva to form a pulp. The nest sites vary according to the species but may be underground, usually in an abandoned small mammal burrow, or aerial under an overhang, in a hollow tree, in a hedge or on a tree branch, or inside a building. The nest consists of combs of hexagonal cells which open downwards, and are surrounded by an envelope. The queen feeds the larvae on macerated insects and spiders. Some species, in subgenus *Paravespula*, are scavengers. The first brood develop into adult workers. These workers build further combs and look after the brood hatching from more queen eggs. Later in the season, combs of larger cells are built in which the new queens are reared. Males are usually reared in the smaller cells; but sometimes also in the large cells. Males and new queens leave the colonies and, after mating, the queens enter overwintering sites in sheltered places *e.g.* under bark and stones, and the males die. The workers gradually all die. One species, *Vespula austriaca*, does not build its own nest but takes over, or usurps, a young colony, usually with just a few workers, of another species, *Vespula rufa*. Such a species is called a social parasite or a cuckoo. The cuckoo queen kills the host queen and uses the workers of the host species to rear a brood of new cuckoo queens and males. The cuckoo species does not produce workers. The queens of non-cuckoo species often attempt to usurp the queen of another colony, usually of the same species but sometimes of a different species.

Diagnosis of British species

All species are social or parasitic species.

Mandibles short, overlapping transversely when fully closed.

Hind wing without a jugal lobe.

Middle tibia with two spurs.

Claws simple.

First gastral segment truncate with anterior surface at right angles to the posterior horizontal surface.

Key to the British species

When mounting specimens it is good practice to expose the posterio-lateral margin of the sixth gastral tergite and sting apparatus of the female and the genitalia of the male. The posterio-lateral margin of the female tergite is displayed by pulling the terminal gastral segments backwards. The male genitalia are large, so easily found and readily pulled out. For a detailed study of the male genitalia it needs to be removed from the body and the left and right sides eased away from each other. Often examination of the male genitalia is the only sure way of determining a species. When the size and density of the punctures on the body need to be observed it may be necessary to scrape off some of the hairs to reveal the punctures.

A key to the world species of the Vespinae is given by Archer (1989a). A check list of the world species in the subfamily Vespinae is given by Carpenter & Kojima (1997). Archer (1989a) also contains a check list of world species although with fewer synonyms than found in Carpenter & Kojima.

The size and colour descriptions for each species should be treated with caution as, particularly for workers, these characteristics can sometimes be very variable.

Key to the genera and subgenera of Vespinae

1. Vertex long, the posterior ocelli more than twice as far from the back of the head as from each other (Fig. 85). Vein Rs reaches vein R+Sc far from the pterostigma; distance from where vein Rs meets vein R+Sc to the pterostigma longer than vein Rs and nearly three times the length of the pterostigma (Fig. 86) .... **Vespa** (p. 53)

- Vertex short, the posterior ocelli about as far from each other as from the back of the head (Fig. 87). Vein Rs reaches vein R+Sc close to the pterostigma; distance from where vein Rs meets vein R+Sc to the pterostigma about equal to or slightly shorter than vein Rs and less than twice the length of the pterostigma (Fig. 88) 2

Figure 85. Dorsal view of head of *Vespa*.

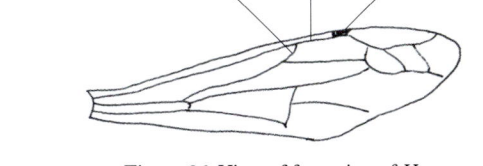

Figure 86. View of fore wing of *Vespa*.

Figure 87. Dorsal view of head of *Vespula* and *Dolichovespula*.

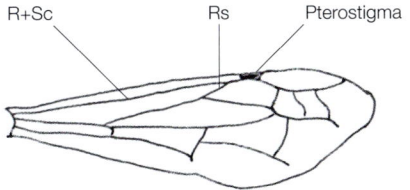

Figure 88. View of fore wing of *Vespula* and *Dolichovespula*.

2. Malar space well developed, longer than the apical diameter of the scape (Fig. 89). Pronotal carina well developed (Fig. 90). (The pronotum can be seen by viewing the thorax from the side) .... *Dolichovespula* (p. 55)

- Malar space short, as short as or shorter than the apical diameter of the scape (Fig. 91). Pronotal carina absent or faintly marked ventrally .... *Vespula* ... 3

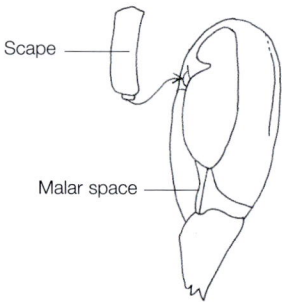
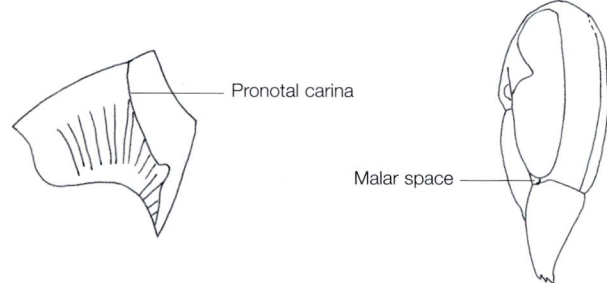

Figure 89. Lateral view of head of *Dolichovespula*.

Figure 90. Lateral view of pronotum of *Dolichovespula media*.

Figure 91. Lateral view of head of *Paravespula*.

3. Long black hairs present on the first gastral tergite. Occipital carina does not extend to the base of the mandible (Fig. 92). Lateral mesoscutal depression without a backwardly-directed depression (Fig. 93). Posterio-lateral margin of the sixth gastral sternite of the female evenly convex (Fig. 97). In the male the profile of the seventh gastral tergite convex (Fig. 94); the shape of the seventh gastral sternite triangular (Fig. 96); the digitus small, not extending to the level of the parameral spine; the apex of the aedeagus strap-shaped (Fig. 95) .... Subgenus *Vespula* s. s. (p. 58)

- Long pale hairs present on the first gastral tergite. Occipital carina extends to the base of the mandible (Fig. 98). Lateral mesoscutal depression with a backwardly-directed depression (sometimes indistinct on workers) (Fig. 99). Posterio-lateral margin of the sixth gastral sternite of the female with a backwardly-directed process (Fig. 103). In the male the profile of the seventh gastral tergite concave (Fig. 100); the shape of the seventh gastral sternite transversely rectangular (Fig. 102); the digitus large, extending nearly as far as or beyond the level of the parameral spine; the apex of the aedeagus not strap-shaped but spoon-shaped (Fig.101) .... Subgenus *Paravespula* (p. 60)

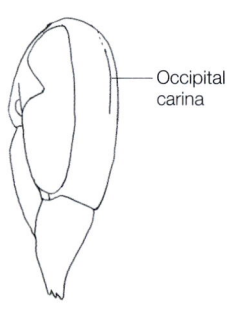
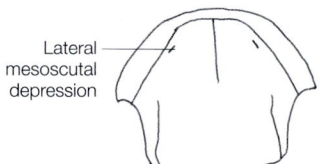

Figure 93. Dorsal view of pronotum and mesoscutum of *Vespula* s.s.

Figure 92. Lateral view of head of *Vespula* s.s.

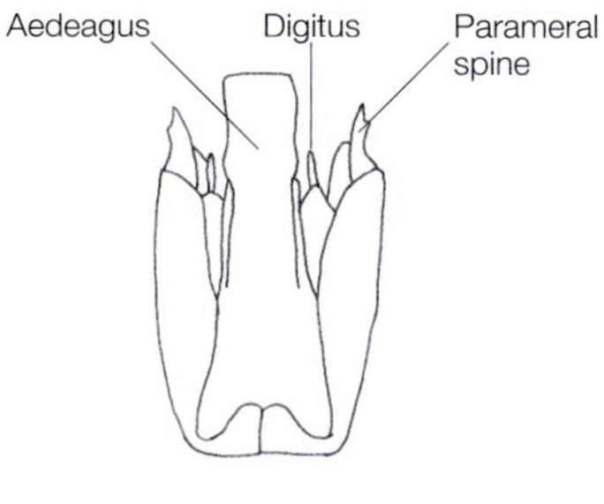

Figure 95. Dorsal view of male genitalia of *Vespula* s.s.

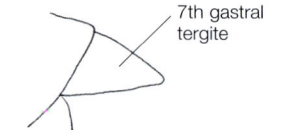

Figure 94. Lateral view of male 7th gastral tergite of *Vespula* s.s.

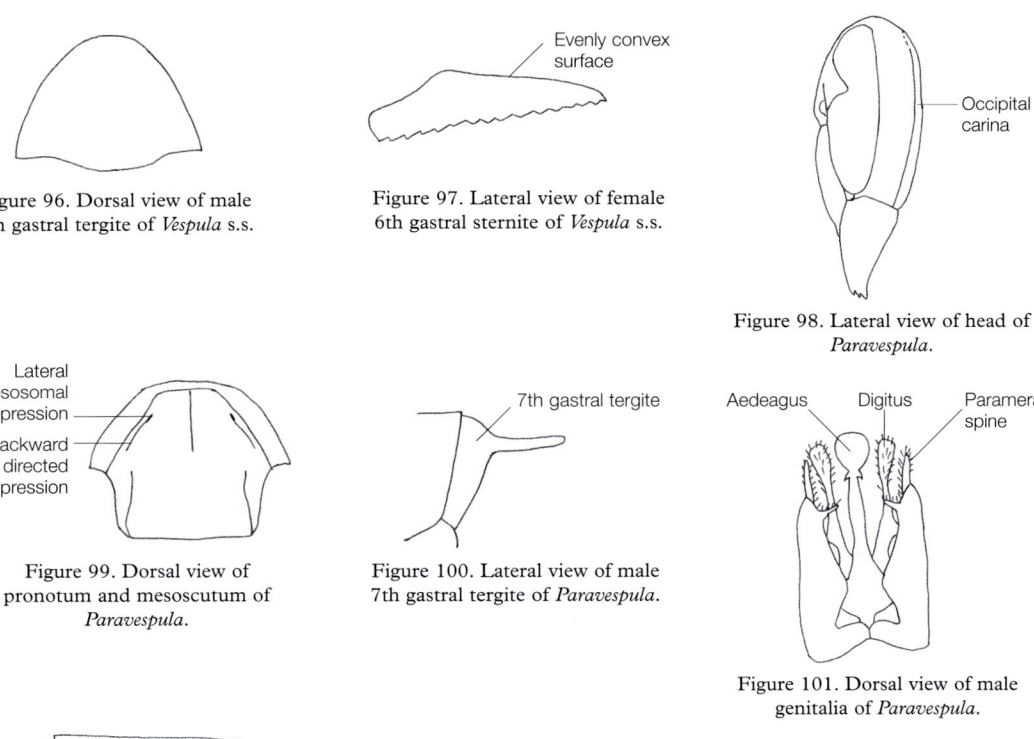

Figure 96. Dorsal view of male
7th gastral tergite of *Vespula* s.s.

Evenly convex surface

Figure 97. Lateral view of female
6th gastral sternite of *Vespula* s.s.

Occipital carina

Figure 98. Lateral view of head of
Paravespula.

Lateral mesosomal depression

Backward directed depression

Figure 99. Dorsal view of
pronotum and mesoscutum of
Paravespula.

7th gastral tergite

Figure 100. Lateral view of male
7th gastral tergite of *Paravespula*.

Aedeagus   Digitus   Parameral spine

Figure 101. Dorsal view of male
genitalia of *Paravespula*.

Figure 102. Dorsal view of male 7th
gastral tergite of *Paravespula*.

Backward-directed process

Figure 103. Lateral view of female
6th gastral sternite of *Paravespula*.

Genus **Vespa** Linnaeus, 1758

This genus consists of 22 species which are mainly restricted to the Oriental region with some reaching the south-east Palaearctic and northern Australasian regions. *V. crabro* occurs in the Palaearctic and Oriental regions and has been introduced into the U.S.A. and Canada. *V. orientalis* Linnaeus, 1771 also occurs in southern Europe and *V. velutina* Lepeletier, 1836 has been accidentally introduced into south-western France and spread to include Brittany bordering the English Channel (Rome *et al.*, 2009). Three species found in Europe and one British species, *V. crabro*.

Key to the species of *Vespa* Linnaeus, 1758

1. Female with the centre of the clypeus covered by such large and coarse punctures that inter-puncture distance is less than puncture diameter and often punctures almost touching each other. Male with same clypeal punctures except less coarse. Vertex yellow. Mesoscutum black with reddish-brown stripes. Mesoscutellum and metanotum reddish brown .... **crabro** (p. 54)

- Female and male with the centre of the clypeus covered with such small punctures that inter-puncture distance is more than puncture diameter. Vertex, mesoscutum, mesoscutellum and metanotum black .... **velutina** (p. 54)

Species accounts

Vespa crabro Linnaeus, 1758 Plate 55

Head and hind two-thirds of gaster mainly yellow, tending to reddish-brown. Mesosoma and first third of gaster mainly reddish-brown sometimes becoming black particularly on the queen. Scape yellow in front. Clypeus and ocular sinus yellow. Gena yellow usually with a variable amount of reddish-brown markings. Mesoscutellum and metanotum reddish brown. Length: queen 26-30 mm, worker 18-23 mm, male 23-24 mm. The hornet is widespread in England and Wales extending from Cornwall to Kent and north to Lancashire and Yorkshire. Also recorded from the Isles of Scilly and the Channel Islands (Jersey and Guernsey). Distribution map in Edwards (1997). Elsewhere, occurs in the Palaearctic and Oriental regions and has been introduced into the U.S.A. and Canada (Archer, 1992). Found in many lowland habitats, but particularly associated with ancient deciduous forests (*e.g.* New Forest and Sherwood Forest) and urban areas. Nests are initiated from the middle of May. Nests are usually situated in aerial situations, particularly inside hollow trees, but also in attics and outhouses. Some nests are built underground. The colony of an underground nest may relocate to another nest site if the initial nesting cavity becomes too small. Queens emerge from overwintering sites from early April. The first workers emerge from late June to early July. Males and new queens mainly emerge during September. Workers may still be seen in October and occasionally early November. Colonies terminate usually during October, or occasionally November. Exudations from damaged roots and branches of oak trees are collected. Ash and lilac twigs are ring-barked to encourage sap flow which is then collected. The mature nest, on average, consists of about 1400 cells (600 small cells and about 800 large cells) in three to eight combs (Archer, 1993, 2008b).

Vespa velutina Lepeletier, 1836

[Not yet recorded in British Isles]

Vertex, mesosoma and antennae, except the underside of the scape, black. The remaining parts of the head reddish-brown. Gastral segments apically black and distally reddish-brown with the reddish-brown colouration increasing in extent on the posterior segments, particularly the tergites. This Asian species was accidentally introduced into Lot-et-Garonne, south west France, probably during 2004. It has rapidly spread across south west France with an outlying record from Brittany, bordering the English Channel, during 2008 (Villemant *et al.*, 2011). The potential invasion risk of this species was assessed using a climatic model which indicated that England, particularly south eastern England, could be climatically suitable (Villemant *et al.*, 2011). In France, the queens probably start nest foundation during May, the first workers appearing during June and the colony terminating during November with the first frosts. By the autumn the colony can have up to 2,000 workers and have produced several hundred new queens and males. The nests are found in urban and rural areas in sheltered situations, *e.g.* in hives, attached to the roof inside a hut, holes in walls, under veranda roof. When nest space is limited so that the nest cannot be expanded, the colony relocates usually to the tops of trees (Perrard *et al.*, 2009; Villemant *et al.*, 2011).

Edwards (1982) reported two queens of *Vespa orientalis* L. in crates of grapefruit from Israel sent to two British supermarkets and also T. Riley (pers. comm., 1981) reported a queen in a crate of jaffa oranges from Israel. The vagrant *V. orientalis* can be readily separated from *V. crabro* and *V. velutina* by colour differences. The gaster of *V. orientalis* is mainly reddish-brown to dark brown with the 3rd, and usually the 4th, gastral tergites yellow.

Genus **Dolichovespula** Rohwer, 1916

This genus consists of 18 species mainly distributed in the Nearctic and Palaearctic regions but also the Oriental region. Seven species are found in Europe, four British species.

Key to the species of *Dolichovespula* Rohwer, 1916

1. The ventral surface of the pronotum is covered with wrinkles (Fig. 105). The indentation or ocular sinus of the eye mainly or entirely yellow (Fig. 104). Male with the seventh gastral sternite strongly notched (Fig. 106) .... *media* (p. 56)

- The ventral surface of the pronotum is without wrinkles. Ocular sinus mainly black. Male with the seventh gastral sternite without a notch ((Fig. 107) 2

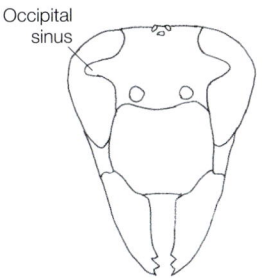
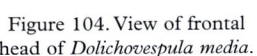

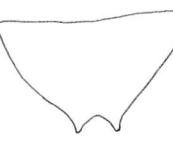
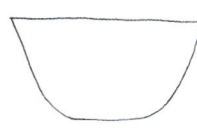

Figure 104. View of frontal head of *Dolichovespula media*.

Figure 105. Lateral view of pronotum of *Dolichovespula media*.

Figure 106. Dorsal view of male 7th gastral sternite of *Dolichovespula media*.

Figure 107. Dorsal view of male 7th gastral sternite of non- *Dolichovespula media* species.

2. Females (Queens & Workers) 3

- Males 5

3. Large punctures more numerous on the clypeus so that generally the distances between punctures equal to or less than puncture diameter. Clypeus yellow or yellow with a small central black spot .... *sylvestris* (p. 57)

- Large punctures less numerous on the clypeus so that the distances between punctures more than puncture diameter. Clypeus yellow with generally a black mark running dorso-ventrally ... 4

4. Head long, POL/PBHL (Post-ocellar line/post-ocellar to back of head line) equal to or less than one (Fig. 108). Hairs on the side of the thorax light brownish. No orange marks on the first and second gastral terga .... *saxonica* (p. 57)

- Head short, POL/PBHL more than one (Fig. 108). Hairs on the side of the thorax black or at least with some black hairs. Orange marks often present on the first and second gastral terga .... *norwegica* (p. 57)

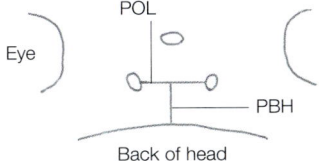

Figure 108. Dorsal view of vertex of head.

5. Dorsal terminal margin of the gonostipes strongly developed as a dorsal terminal process and directed backwards. Digitus short, not extending beyond the level of the aedeagus (Fig. 109) .... *sylvestris* (p. 57)

- Dorsal terminal margin of the gonostipes weakly developed as a dorsal terminal process and not directed backwards (Fig. 110). Digitus large, extending to or beyond the level of the aedeagus 6

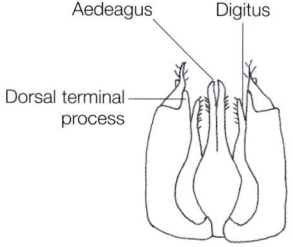

Figure 109. Dorsal view of male genitalia of *Dolichovespula sylvestris*.

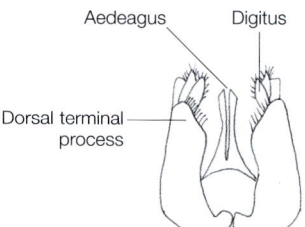

Figure 110. Dorsal view of male genitalia of *Dolichovespula norwegica*.

6. Dorsal terminal margin of the gonostipes only projecting slightly inwards (Fig. 110). Apical antennal segments with a swelling or tyloid each. Usually specimens have orange marks on the second and often the third gastral tergites. Clypeal black stripe usually complete being connected to the dorsal and ventral margins .... *norwegica* (p. 57)

- Dorsal terminal margin of the gonostipes strongly projecting inwards (Fig. 111). Apical antennal segments with one or two tyloids. No orange marks on the second and third gastral tergites. Clypeal black stripe often incomplete being absent dorsally ..... *saxonica* (p. 57)

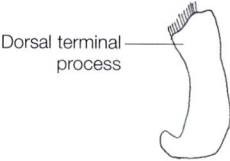

Figure 111. Dorsal view of male genitalia of *Dolichovespula saxonica*.

Species accounts

Dolichovespula media (Retzius, 1783) Plate 51

Yellow and black, queen may have reddish brown makings. Scape yellow in front, may be reddish-brown on queen. Workers may have mainly black gaster with a very narrow yellow band on each tergite. Ocular sinus yellow. Clypeus yellow with or without an incomplete dorso-ventral black stripe. Gena yellow. Queen with mesoscutellum largely reddish-brown and metanotum with reddish-brown marks. Worker and male with lateral yellow marks on the mesoscutellum and metanotum which may be absent in dark forms. Length: queen 17-21 mm, worker 14-17 mm, male 15-20 mm. First recorded in East Sussex during 1980, it is now distributed from Cornwall to Kent and north to North Wales, Cumbria, Yorkshire and southern Scotland. Distribution map in Edwards (1997). Elsewhere, it occurs in northern and central Europe, North Africa and across northern and central Asia from Turkey to Japan (Archer, 1999). Found in lowland habitats including urban areas. Nests are built in aerial situations, usually suspended from branches of trees and shrubs, from ground level to a height of several metres. A few nests have been found attached to walls of houses, inside a lamp bracket and beneath the eaves of a caravan. Nests are initiated from early May, with the first workers appearing from early June and the first males and new queens from early August. Colonies terminate by the end of August. The nest, on average, consists of about 1000 cells (300 small cells and about 700 large cells) in three to five combs (Archer, 2006).

Dolichovespula norwegica (Fabricius, 1781) Plate 52

Yellow and black, sometimes with lateral reddish markings on tergites one to three. Scape yellow in front. Ocular sinus black with a narrow ventral yellow stripe. Clypeus yellow with a broad complete dorso-ventral black stripe, stripe narrower on male. Gena black with a dorsal yellow marking. Mesoscutellum with lateral yellow markings, metanotum black. Length: queen 15-17 mm, worker 12-14 mm, male 13-15 mm. Found throughout mainland England, Wales, Scotland and Ireland including Orkney and Shetlands. Distribution map in Edwards & Telfer (2002). Elsewhere, it occurs in Europe (rare in southern Europe), North America and northern and central Asia from Turkey to Sakhalin (Archer, 1999). Found in many open habitats including urban areas but perhaps with a preference for heathland and moorland. Usually nests are made in aerial situations attached to shrubs from ground level to about one metre above the ground and on trees up to ten metres above the ground. Also nests have been found under eaves of a house, on a wall, attached to a window frame, in a beehive and under a garden seat. Queens emerge from overwintering sites from late April with nest foundation from mid-May. The first workers emerge from early June, the new queens and males from mid-July and the colony terminates by the end of August. The mature nest, on average, consists of about 1650 cells (350 small cells and about 1300 large cells) in usually three to five combs (Archer, 2000a, 2006). Each colony either rears mainly males or new queens.

Dolichovespula saxonica (Fabricius, 1793) Plate 53

Yellow and black. Scape yellow in front. Ocular sinus black with a narrow ventral yellow stripe. Clypeus yellow with a variable sized dorso-ventral stripe which is often incomplete and much reduced. Gena black with dorsal and ventral markings. Mesoscutellum with lateral yellow markings, metanotum black or may have small lateral yellow marks. Length: queen 15-18 mm, worker 11-15 mm, male 12-17 mm. First recorded in Surrey during 1987, it is now distributed from Devon to Cornwall and northwards to Lancashire and Yorkshire and Wales. Distribution map in Edwards (1997). Elsewhere, it occurs in central and northern Europe, central and northern Asia from Turkey to Japan (Archer, 1999). Found in many open habitats usually nesting in aerial situations such as shrubs and trees up to 2 metres above the ground, in beehives, under the eaves of buildings up to seven metres above the ground, under porches, in wall cavities and rarely in a tree hole or a hole in the ground. Probably nest foundation starts just before mid-May with the first workers emerging during early June. The new queen and males emerge by early July with colony termination soon after mid-August. The mature nest, on average, consists of about 1330 cells (430 small cells and 900 large cells) in three to five combs (Archer, 2006).

Dolichovespula sylvestris (Scopoli, 1763) Plates 54 & 61

Yellow and black. Scape yellow in front. Ocular sinus ventrally yellow. Clypeus yellow or with a central black spot, male with an incomplete dorso-ventral stripe. Gena black and yellow. Mesoscutellum with lateral yellow markings, metanotum black. Length: queen

16-19 mm, worker 13-15 mm, male 14-17 mm. Found throughout mainland England, Wales, Scotland and Ireland including the Isle of Man, Isles of Scilly, the Channel Islands and the Outer Hebrides. Distribution map in Edwards & Telfer (2002). Elsewhere, it occurs in Europe, North Africa and northern and central Asia from Turkey to eastern China (Archer, 1999). Usually nests are in, or partially in, enclosed aerial sites such as under overhanging banks, in bird boxes, under eaves and porches, in outhouses, roof spaces and beehives. Nests are also found in shallow holes in the ground when the nest is usually visible at the soil/leaf litter interface. The overwintering queens emerge from late April with nest foundation from mid-May. The first workers emerge from early June, the new queens and males from mid-July and the colony usually terminates by the end of August. The mature nest, on average, has about 800-900 cells (300 small cells, 500-600 large cells) in three to four combs Archer (1981, 2002, 2006).

Genus **Vespula** Thomson, 1869

Subgenus **Vespula** s. s.

This subgenus consists of ten species mainly distributed in the Nearctic and Palaearctic regions but also the Oriental region and just into the Neotropical region. Two species found in Britain and Europe.

Key to the species of subgenus *Vespula* s. s.

1. Hind tibia with long erect hairs (Fig. 113). Anterior angles of the clypeus sharply produced in females, bluntly produced in males (Fig. 112). Worker caste absent ....
.... ***austriaca***

- Hind tibia without long erect hairs. Anterior angles of the clypeus bluntly produced in females, hardly produced at all in males (Fig. 112). Worker caste present .... ***rufa***

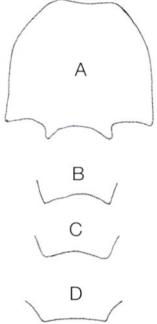

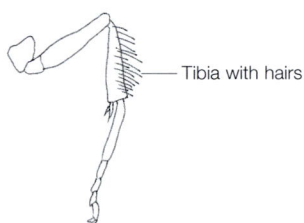

Figure 113. Hind leg of *Vespula austriaca*.

Figure 112. Clypeus of *Vespula austriaca* queen (A),
ventral clypeal margins of *V. austriaca* male (B), of *V. rufa*
queen and worker (C) and *V. rufa* male (D).

Species accounts

Vespula austriaca (Panzer, 1799) Plate 56

Yellow and black. Scape black in front, male yellow in front. Ocular sinus black with a narrow ventral yellow stripe. Clypeus yellow usually with a small three-pronged black mark. Gena black with a dorsal yellow mark, male usually with complete yellow stripe. Mesoscutellum with lateral yellow markings, metanotum black. Length: queen 16-18 mm, male 12-15 mm. Found throughout Britain and Ireland but with few records from southeast England. Distribution map in Edwards (1998). Elsewhere, found in central and northern Europe, Asia (Turkey to northern Pakistan, northern China, Japan), Canada and northern U.S.A. (Archer, 1997a). Found in open habitats where the nests of its host (*V. rufa*) are found. The overwintering queens emerge from late May until early July, but usually during June. New sexuals emerge from mid- to late July, mating from early August. The queen enters the host colony after the first *V. rufa* workers have emerged usually when there are at least ten workers present, and immediately kills the host queen or drives her away from the nest. The workers resist the usurping queen but she dominates them, probably physically. The queen lays her eggs in the *V. rufa* cells and these are then reared by the host workers. The *V. rufa-austriaca* nests at maturity are smaller than those of the pure *V. rufa*. On average, the nests consist of just over 500 cells (200 large cells and just over 300 small cells) in two or three combs. On average, each colony probably rears about 100 queens and 100 males of *V. austriaca* and about 200 workers of *V. rufa* (Archer, 1997a, 2007).

Vespula rufa (Linnaeus, 1758) Plate 58

Yellow and black often with reddish areas on the first two tergites. Scape black in front, male yellow in front. Ocular sinus black with a narrow ventral yellow stripe. Clypeus with an incomplete dorso-ventral black marking, male yellow or more usually with a small black spot. Gena black with a dorsal yellow marking and sometimes a smaller ventral yellow marking, male a continuous yellow stripe. Mesoscutellum with lateral yellow markings, metanotum black or with small lateral yellow markings. Length: queen 17-18 mm, worker 12-14 mm, male 13-15 mm. Found throughout mainland England, Wales, Scotland and Ireland. Also recorded from the Isle of Man, the Channel Islands and the Outer Hebrides. Distribution map in Edwards & Telfer (2001). Elsewhere, it is found in central and northern Europe besides being distributed across northern and central Asia from Turkey to Japan and North America (Archer, 1997a). Found in open habitats, e.g. open woodlands, moorland and hedge banks, less commonly in urban situations. Nests are usually underground in a dry, often shaded situation, close to the soil surface, even just under the leaf litter or moss layer. Also in hollow tree stumps and suspended from the roots of trees in hollows in the ground. Nests are rarely above ground (in dense bushes, cavity walls, bird boxes and eaves of houses). The overwintering queens emerge from late March until early May with the first workers emerging from the end of May until mid-June. New sexuals emerge from late July; mating usually from early August. The colony terminates during September. The mature nest, on average, consists of about 800-850 cells (400 small cells and 400-450 large cells) in three combs (Archer, 1997b, 2007).

Subgenus **Paravespula** Blüthgen, 1938

The subgenus consists of eleven species distributed in the Palaearctic, Nearctic and Oriental regions, although *V. vulgaris* and *V. germanica* are now cosmopolitan in distribution and *V. pennsylvanica* (de Saussure, 1857) has been introduced into the Hawaiian Islands. Two species are found in Britain and Europe.

Key to the species of the subgenus *Paravespula* Blüthgen, 1938

1. Female (queens and workers) with the margin behind the third mandibular tooth straight or at most slightly concave (Fig. 115); yellow genal band interrupted by a black bar which may be reduced to a black spot. Aedeagus of male with a small backwardly directed barb on each side below the apical spooned-shaped region. Aedeagus rounded apically (Fig. 114A) .... *vulgaris*

- Female (queens and workers) with the margin behind the third mandibular tooth distinctly concave (Fig. 115); yellow genal band not interrupted by a black bar or spot Aedeagus of male without a small backwardly directed pointed barb on each side below the apical spooned-shaped region. Aedeagus concave apically (Fig. 114B) .... *germanica*

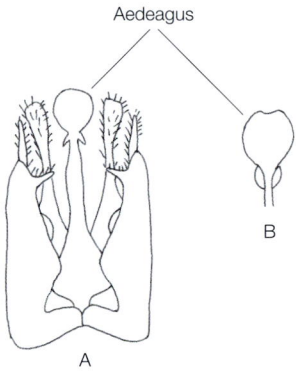

Figure 114. Dorsal view of male genitalia of *Vespula vulgaris* (A) and aedeagus of *V. germanica* (B)

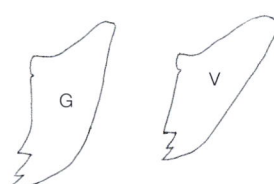

Figure 115. View of female mandible of *Vespula germanica* (G) and *V. vulgaris* (V).

Species accounts

Vespula germanica (Fabricius, 1793) Plate 57

Yellow and black. Scape black in front, male yellow in front. Ocular sinus yellow with inner margin convex. Clypeus yellow usually with a reduced incomplete dorso-ventral stripe and black spots or just black spots. Gena yellow. Mesoscutellum and metanotum with lateral yellow markings. Length: queen 16-19 mm, worker 12-14 mm, male 14-17 mm. Found throughout mainland England, Wales, Scotland and Ireland. It is less common in north-west Scotland. Also recorded from the Isles of Scilly, the Channel Islands and Shetland. Distribution map in Edwards & Telfer (2002). Elsewhere, it occurs in Europe, North Africa, and northern and central Asia from Turkey to China. Introduced to New Zealand, Tasmania, southern Australia, Ascension Islands, South Africa, Chile, Argentina, northern U.S.A., south to California, and Canada (Archer, 1998b). Found in many, usually open, habitats including urban areas. Nest sites are mostly in deep (about

10 cm) underground sites (*e.g.* old rodent burrows). Aerial sites are enclosed and include roof spaces, beehives or thick hedges. Aerial sites are more frequent in urban areas. Overwintering queens emerge from mid-March with nest foundation from early May. The first workers emerge from early June, the new queens and males from early September and the colony terminates early November and rarely mid-December. Mature nests, on average, consist of about 7600-8000 cells (6100-6500 small cells and 1500 large cells) in eight to nine combs (Archer, 2001, 2008a).

Vespula vulgaris (Linnaeus, 1758) Plate 59

Yellow and black. Scape black in front, male yellow in front. Ocular sinus yellow with inner margin concave. Clypeus yellow with an incomplete dorso-ventral black mark, narrower on male. Gena yellow interrupted by a black mark, yellow on male. Mesoscutellum and metanotum with lateral yellow markings. Length: queen 15-18 mm, worker 10-13 mm, male 12-17 mm. Found throughout mainland England, Wales, Scotland and Ireland. Also recorded from the Isle of Man, Isles of Scilly, the Channel Islands, Outer Hebrides, Orkneys and Shetland. Distribution map in Edwards & Telfer (2002). Elsewhere, it occurs in Europe, North Africa, northern and central Asia from Turkey to Japan and introduced into Iceland, New Zealand and south-eastern Australia (Archer, 1998b). Found in many usually open habitats including urban areas. Nest sites are mostly in deep (8-15 cm) underground sites (e.g. old rodent burrows). Aerial sites are enclosed and include outhouses, cavity walls, roof spaces, beehives, tree hollows and dense vegetation. Aerial sites are more frequent in urban areas. Overwintering queens emerge from early March with nest foundation from mid-May. The first workers emerge from early June, the new queens and males from mid-September and the colony terminates usually from the end of October and rarely to early February. Mature nests, on average, consist of about 8250-8700 cells (6500 small, 1750-2200 large) in 6-12 combs (Archer, 1981, 2003b, 2008a).

References

(Most of the unpublished are available on the BWARS website – www.bwars.com)

Akre, R.D., Greene, A., MacDonald, J.F., Landolt, P.J. and Davis, H.G. 1980. *Yellowjackets of America North of Mexico.* U.S. Department of Agriculture, Agriculture Handbook No. 552, 102pp.

Archer, M.E. 1977. The weights of forager load of *Paravespula vulgaris* (Linn.) (Hymenoptera: Vespidae) and the relationship of load weight to forager size. *Insectes Sociaux* **24**: 95-101.

Archer, M.E. (2nd ed.). 1979. *Provisional atlas of the Insects of the British Isles. Hymenoptera: Vespinae. Social Wasps.* Biological Records Centre, Huntingdon. 18pp.

Archer, M.E. 1981. A simulation model for the colonial development of *Paravespula vulgaris* (Linnaeus) and *Dolichovespula sylvestris* (Scopoli) (Hymenoptera: Vespidae). *Melanderia* **36**: 1-59.

Archer, M.E. 1989a. A key to the World Species of the Vespinae (Hymenoptera). Part 1 Keys, Checklist and Distribution. Part 2 Figures. *Research Monograph of the College of Ripon and York St. John* **2**: 1-47, 1-34.

Archer, M.E. 1989b. *Odynerus simillimus* Morawitz (Hym., Eumenidae) taken in Norfolk. *Entomologist's Monthly Magazine* **125**: 206.

Archer, M.E. 1992. The taxonomy of *Vespa crabro* L. and *V. dybowskii* André (Hym., Vespinae). *Entomologist's Monthly Magazine* **128**: 157-163.

Archer, M.E. 1993. The life-history and colonial characteristics of the Hornet, *Vespa crabro* L. (Hym., Vespinae). *Entomologist's Monthly Magazine* **129**: 151-163.

Archer, M.E. 1997a. Taxonomy, distribution and nesting biology of the species of the Euro-Asian *Vespula rufa* group (Hym., Vespinae). *Entomologist's Monthly Magazine* **133**: 107-114.

Archer, M.E. 1997b. A numerical account of successful colonies of the social wasp, *Vespula rufa* (L.) (Hym., Vespinae). *Entomologist's Monthly Magazine* **133**: 205-215.

Archer, M.E. 1998a. *Rhynchium oculatum* (F.) (Hym., Eumenidae) found at Nottingham, England. *Entomologist's Monthly Magazine* **134**: 263.

Archer, M.E. 1998b. The world distribution of the Euro-Asian species of *Paravespula* (Hym., Vespinae). *Entomologist's Monthly Magazine* **134**: 279 284.

Archer, M.E. 1999. Taxonomy and world distribution of the Euro-Asian species of *Dolichovespula* (Hym., Vespinae). *Entomologist's Monthly Magazine* **135**: 153-160.

Archer, M.E. 2000a. The life history and a numerical account of colonies of the social wasp, *Dolichovespula norwegica* (F.) (Hym., Vespinae) in England. *Entomologist's Monthly Magazine* **136**: 1-14.

Archer, M.E. 2000b. *Eumenes papillarius* (Christ) (Hym., Vespinae) recorded again from Yorkshire. *Entomologist's Monthly Magazine* **136**: 210.

Archer, M.E. 2001. A numerical account of the development of colonies of the social wasp, *Paravespula germanica* (F.) (Hym., Vespinae) in England. *Entomologist's Monthly Magazine* **137**: 115-134.

Archer, M.E. 2002. A numerical account of the development of colonies of the social wasp, *Dolichovespula sylvestris* (Scopoli) (Hym., Vespinae), in England. *Entomologist's Monthly Magazine* **138**: 209-223.

Archer, M.E. 2003a. (2nd. ed.). *The British Potter and Mason Wasps. A Handbook.* Vespid Studies, York, 96pp.

Archer, M.E. 2003b. A numerical account of the development of successful colonies of the social wasp, *Paravespula vulgaris* (L.) (Hym., Vespinae) in England and overseas. *Entomologist's Monthly Magazine* **139**: 139-160.

Archer, M.E. (ed.). 2004. *BWARS. Bees, Wasps and Ants Recording Society. Members' Handbook.* Centre for Ecology and Hydrology, Huntingdon, 155pp.

Archer, M.E. 2005. A numerical model of seasonal foraging characteristics of successful underground colonies of *Vespula vulgaris* (Hymenoptera, Vespidae) in England. *Insectes Sociaux* **52**: 231-237.

Archer, M.E. 2006. Taxonomy, distribution and nesting biology of species of the genus *Dolichovespula* (Hymenoptera, Vespidae). *Entomological Science* **9**: 281-293.

Archer, M.E. 2007. Taxonomy, distribution and nesting biology of species of the genus *Vespula* or the *Vespula rufa* species group (Hymenoptera: Vespidae). *Entomologist's Monthly Magazine* **143**: 35-48.

Archer, M.E. 2008a. Taxonomy, distribution and nesting biology of species of the genus *Paravespula* or the *Vespula vulgaris* species group (Hymenoptera: Vespidae). *Entomologist's Monthly Magazine* **144**: 5-29.

Archer, M.E. 2008b. Taxonomy, distribution and nesting biology of species of the genera *Provespa* and *Vespa* (Hymenoptera: Vespidae). *Entomologist's Monthly Magazine* **144**: 69-101.

Baldock, D.W. 2003. Wildlife Reports. Bees, wasps and ants. *British Wildlife* **15**: 63-64.

Baldock, D.W. 2004. Wildlife Reports. Bees, wasps and ants. *British Wildlife* **16**: 60-62.

Baldock, D.W. 2005. Wildlife Reports. Bees, wasps and ants. *British Wildlife* **17**: 57-59.

Barrington, R. and Edwards, M. 1991. Observations of nest building by *Eumenes coarctatus* (L.) (Hym., Eumenidae) on a West Sussex heathland. *Entomologist's Monthly Magazine* **127**: 118.

Bignell, G.C. 1882. *Odynerus pictus* Curt. *Entomologist* **15**: 164.

Blair, K.G. 1943. A nest of *Ancistrocerus pictus* (Hym., Vespidae). *Entomologist's Monthly Magazine* **79**: 184.

Blüthgen, P. 1961. Die Faltenwespen Mitteleuropas (Hymenoptera, Diploptera). *Abhandlungen Der Deutschen Akademie der Wissenschaften zu Berlin (Klasse für Cheie., Geologie und Biologie)*. **2**: 1-248.

Bridgman, J.B. 1887. *Eumenes coarctata* and its parasite. *The Entomologist* **20**: 18-19.

Carpenter, J.M. 1996. Distributional checklist of species of the genus *Polistes* (Hymenoptera: Vespidae: Polistinae, Polistini). *American Museum Novitates* **3188**:1-39.

Carpenter, J.M. and Kojima, J. 1997. Checklist of the species of the subfamily Vespinae (Insecta: Hymenoptera: Vespidae). *Natural History Bulletin of Ibaraki University*. **1**: 51-92.

Chambers, V.H. 1944. Nidification of *Odynerus laevipes* Shuck. (Hym., Vespidae). *Entomologist's Monthly Magazine* **80**: 11.

Chambers, V.H. 1949. The Hymenoptera aculeata of Bedfordshire. *Transactions of the Society for British Entomology* **9**: 197-252.

Chapman, T.A. 1870. Note on the pairing of *Odynerus spinipes*, Linn. *Entomologist's Monthly Magazine* **6**: 214.

Chapman, T.A. 1877. On the chrysides parasitic on *Odynerus spinipes*. *Transactions of the Cardiff Naturalists' Society* **9**: 91-100.

Clark, J. 1924. The Bees, Wasps and Ants of Roxburghshire. *Transactions of the Haworth Archaeological Society* **1924**: 13-20.

Clemons, L. 1993. *Polistes dominulus* (Christ, 1791) (Hymenoptera: Vespidae) in Greater London. *Entomologist's Record* **105**: 177.

Cooter, J. 1976. *Polistes metricus* (Say) (Hym., Vespidae) accidentally imported into Britain. *Entomologist's Monthly Magazine* **112**: 122.

Danks, H.V. 1971. Biology of some stem-nesting aculeate Hymenoptera. *Transactions of the Royal Entomological Society of London* **122**: 323-399.

Dieck, C. 2008. *Pseudepipona herrichii*. *Hymettus Report* **2008**: 6-7 (unpublished).

Dieck, C. 2010. *Pseudepipona herrichii*. *Hymettus Report* **2010**: 10-11 (unpublished).

Dieck, C. and Neal, R. 2007. *Pseudepipona herrichii*. *Hymettus Report* **2007**: 19-20 (unpublished).

Dvořák, L. and Roberts, P.M. 2006. Key to the paper and social wasps of Central Europe (Hymenoptera: Vespidae). *Acta Entomologica Musei Nationalis Pragae* **46**: 221-244.

Edwards, M. 2007. Observations of prey capture in *Odynerus spinipes* and *Odynerus melanocephalus* at Porth Neigie' Lleyn Peninsula, North Wales, 24-26 May 2007. *Hymettus Report* **2007**: 15-16 (unpublished).

Edwards, M. and Roberts, S.P.M. 1995. *The current status of Pseudepipona herrichii (Saussure) (Hymenoptera Aculeata: Eumenidae) at Godlington Heath NNR Dorset, with studies of its autecology.* English Nature, Pre-recovery project: 1-20 (unpublished).

Edwards, R. 1980. *Social Wasps. Their biology and control.* The Rentokil Library, 398pp.

Edwards, R. 1982. Travelling Hornets. *Sphecos* **2**: 9.

Edwards, R. (ed.). 1997. *Provisional atlas of the aculeate Hymenoptera of Britain and Ireland. Part 1.* Biological Records Centre, Huntingdon.

Edwards, R. (ed.). 1998. *Provisional atlas of the aculeate Hymenoptera of Britain and Ireland. Part 2.* Biological Records Centre, Huntingdon.

Edwards, R. and Telfer, M.G. (eds). 2001. *Provisional atlas of the aculeate Hymenoptera of Britain and Ireland. Part 3.* Biological Records Centre, Huntingdon.

Edwards, R. and Telfer, M.G. (eds). 2002. *Provisional atlas of the aculeate Hymenoptera of Britain and Ireland. Part 4.* Biological Records Centre, Huntingdon.

Else, G.R. 1973. Hymenoptera Aculeata, *in*, Appleton, D., Dickson, R. and Else, G. *The Insects of Oxenbourne Down*: 65-107. Fareham.

Else, G.R. 1992. Two little-known wasps: *Homonotus sanguinolentus* (F.) and *Euodynerus quadrifasciatus* (F.) (Hym., Aculeata) in southern England in 1990 and a review of their occurrence in Britain. *Entomologist's Monthly Magazine* **128**: 67-68.

Else, G.R. 2006. Provisional observations on the autecology of *Euodynerus quadrifasciatus* (Fabricius, 1793) (Hymenoptera, Vespidae, Eumeninae) at the Isle of Portland, Dorset, June 2006. *Hymettus Report* **2006**: 17-22 (unpublished).

Else, G.R. and Roberts, S.P.M. 2006. Provisional observations on the autecology of *Euodynerus quadrifasciatus* (Fabricius, 1793) (Hymenoptera, Vespidae, Eumeninae) at the Isle of Portland, Dorset, June 2006. *BWARS Newsletter* **2006 (Autumn)**: 13-18.

Falk, S. 1991. A review of the scarce and threatened bees, wasps and ants of Great Britain. *Research and Survey in Nature conservation* **35**: 1-344.

Falk, S. 2005. A study of the Mason Wasp *Odynerus melanocephalus* in Warwickshire. *ACG Report* **2005**: (unpublished).

Gauld, I. and Bolton, B. (eds). 1996. *The Hymenoptera.* Reprinted by Oxford University Press, England, xi + 332pp.

Goulet, H and Huber, J.T. (eds). 1993. *Hymenoptera of the world: An identification guide to families.* Research Branch. Agriculture. Canada, 668pp.

Guichard, K.M. 1972. *Symmorphus crassicornis* (Panzer) (Hym., Vespoidea) in Britain, with a key to the British species of *Symmorphus* Wesmael. *Entomologist's Gazette* **23**: 169-173.

Guichard, K.M. 1991. The occurrence of *Eumenes papillarius* (Christ) (Hym., Vespidae) in England. *Entomologist's Monthly Magazine* **127**: 71.

Harvey, P.R. 2001. *Odynerus simillimus. Aculeate Study Group* **2000**: 10-14 (unpublished).

Harvey, P.R. 2002. *Odynerus simillimus. Aculeate Study Group* **2002**: 8-11 (unpublished).

Harvey, P. and Scott, D. 2001. *Odynerus simillimus* Morawitz, F., 1867, rediscovered in Essex. *Entomologist's Monthly Magazine* **137**: 226.

Hunnisett, J. 2006. *Odynerus melanocephalus. Hymettus Report* **2006**: 23-29 (unpublished).

Jones, H.P. 1937. A new British Mason Wasp. *Entomologist's Monthly Magazine* **73**: 13-15.

Jørgensen, P. 1942. Biological observations on some solitary vespides. *Entomologiske Meddelelser* **22**: 299-335.

Julliard, C. 1950. Nid de l'*Odynerus scoticus* Curtis (Hym., Vespidae). *Bulletin de l a Société Entomologique Suisse* **23**: 369-376.

Knowles, A. 2006. The paper wasp *Polistes dominulus* new to Essex. *BWARS Newsletter* **2006 (Autumn)**: 28.

Kunz, P.X. 1994. Die Goldwespen (Chrysididae) Baden-Württembergs. *Beihefte zu den Veröffentlichungen für Naturschutz und Landschaftplege in Baden-Württemberg* **77**: 1-188.

Lee, P. and Scott, D. 2007. East Anglian wetland species. *Hymettus Ltd. Report* **2007**: 35-37.

Lucas, W.J. 1931. Mud cells (nests) of *Eumenes coartata*. *The Entomologist* **64**: 144-146.

McClenaghan, I. 1979. *Polistes omissus* Weyrauch (Hymenoptera: Vespidae) in Bangor, Co. Down: a species new to the British list. *Irish Naturalist Journal* **19**: 448.

McKinley, R.C. 1980. *Polistes (Polistes) gallicus* (L.) (Hym., Vespidae) in Scotland. *Entomologist's Monthly Magazine* **116**: 11.

Morgan, D. 1984. Cuckoo-Wasps. Hymenoptera, Chrysididae. *Handbooks for the Identification of British Insects.* **6 (5)**: 1-37.

Morice, F.D. 1906. Nidification of *Odynerus reniformis* Gmel., near Chobham. *Entomologist's Monthly Magazine* **42**: 216-220.

Morrison, S. 1991. The status, distribution and conservation of *Pseudepipona herrichii*. *Report to the Nature Conservancy Council* **1991**: 1-14 (unpublished).

Neal, R. 2006. *Pseudepipona herrichii*. *Hymettus Report* **2006**: 30-33 (unpublished).

Nielsen, E.T. 1932. Sur les habitudes des Hyménoptéres aculeates. II. Vespidae, Chrysididae, Sapygidae et Mutillidae. *Entomologiske Meddelelser* **18**: 84-174.

Nielsen, E.T. 1942. Notes sur le biologie de *Pterocheilus* Klug. *Entomologiske Meddelelser* **22**: 290-294.

Perrard, A., Haxaire, J., Rortais, A. & Villemant, C. 2009. Observations on the colony activity of the Asian hornet *Vespa velutina* Lepeletier, 1836 (Hymenoptera: Vespidae: Vespinae) in France. *Annales de la Société entomologique de France.* **45**: 119-127.

Proctor, M. and Yeo, P. 1973. *The Pollination of Flowers*. Collins, London, 418pp.

Proctor, M., Yeo, P. and Lack, A. 1996. *The Natural History of Pollination*. Collonis, London, 479pp.

Ratzeburg, J.T.C. 1852. Die Ichneumomen der Forstinsecten in forstlicher und entomologischer Beziebung, *ein*, Anhang zur Abbildung und Beschreibung der Forstinsecten. Vol. III, Berlin.

Reeve, H.K. 1991. *Polistes*, *in*, Ross, K.G. and Matthews, R.W. (eds) *The Social Biology of Wasps*: 99-148. Cornell University Press.

Richards, O.W. 1958. Two rare species of *Odynerus* s.l. (Hym., Vespidae). *Entomologist's Monthly Magazine* **94**: 192.

Richards, O.W. 1978. *The social wasps of the Americas*. British Museum (Natural History), London, 580pp.

Richards, O.W. 1980. Scolioidea, Vespoidea and Sphecoidea. Hymenoptera, Aculeata. *Handbooks for the Identification of British Insects.* **6(3b)**: 1-118.

Roberts, S.P.M. 1998. Species Recovery Program. Purbeck Mason-wasp; *Pseudepipona herrichii* (Saussure) (Hymenoptera Aculeata: Eumenidae). *English Nature* **1998**: 1-38 (unpublished).

Roberts, S.P.M. 1999. Species Recovery Program. Purbeck Mason-wasp; *Pseudepipona herrichii* (Saussure) (Hymenoptera Aculeata: Eumenidae). *English Nature* **1999**: 1-36 (unpublished).

Roberts, S.P.M. 2000. Species Recovery Program. Purbeck Mason-wasp; *Pseudepipona herrichii* (Saussure) (Hymenoptera Aculeata: Eumenidae). *English Nature* **2000**: 1-36 (unpublished).

Roberts, S.P.M. 2001. Species Recovery Program. Purbeck Mason-wasp; *Pseudepipona herrichii* (Saussure) (Hymenoptera Aculeata: Eumenidae). *English Nature* **2001**: 1-44 (unpublished).

Roberts, S.P.M. 2008. *Pseudepipona herrichii. Hymettus Report* **2008**: 1-11 (unpublished).

Roberts, S.P.M. and Else, G.R. 1997. The Purbeck Mason-wasp – back from the brink? *British Wildlife* **9**: 90-94.

Rome, Q., Muller, F., Gargominy O. & Villemant, C. 2009. Bilan 2008 de l'invasion de *Vespa velutina* Lepeletier en France (Hymenoptera, Vespidae). *Bulletin de la Société entomologique de France* **114**: 297-302.

Saunders, E. 1896. *The Hymenoptera Aculeata of the British Islands.* Reeve, London, xii+391pp.

Saunders, E. 1903. *Odynerus (Hoplopus) simillimus*, Mor., a wasp new to the British list. *Entomologist's Monthly Magazine* **39**: 6-7.

Schneider, N. 1991. Contribution à la connaissance des *Arthropodes rubicoles* du Grand Duché de Luxembourg. *Bulletin de la Société naturelle de Luxembourg* **92**: 85-119.

Schneider, N. and Leclercq, J. 1987. Nidification d'une guêpe solitaire (Hym. Eumenidae) dans un rayon d'une Abeille sociale (Hym. Apidae). *L'Entomologiste* **43**: 269-270.

Scott, D. 2003. 2002 update of *Odynerus simillimus* in north-east Essex. *Essex Naturalist* **20**: 52-53.

Scott, D. & Strudwick, T. 2009. *Odynerus simillimus. Hymettus Report* **2009**: 5-8 (unpublished).

Sheppard, T. 1926. Mason wasp cells in lock. *Naturalist* **15**: 268.

Shirt, D.B. (ed.). 1987. *British Red Data Books: 2. Insects.* Nature Conservancy Council, Peterborough, 402pp.

Shuckard, W.W. 1837. *Essay on the Indigenous Fossorial Hymenoptera.* Shuckard, London, 259pp.

Smith, F. 1858. *Catalogue of British Fossorial Hymenoptera, Formicidae, and Vespidae, in the collection of the British Museum.* British Museum, London, 236pp.

Smith, F. 1875-6. On the capture of a South American wasp near Liverpool. *Entomologist's Monthly Magazine* **12**: 171-172.

Smith, F. 1878. Entomological Echoes. *The Entomologist* **11**: 156-157.

Spooner, G.M. 1934. Observations on *Odynerus (Lionotus) herrichii* Sauss. In Dorset. *Entomologist's Monthly Magazine* **70**: 46-54.

Spradbery, J.P. 1973. *Wasps. An account of the biology and natural history of solitary and social wasps.* Sidgwick and Jackson, London, 408pp.

Stelfox, A.W. 1933. Some recent records of Irish Aculeate Hymenoptera. *Entomologist's Monthly Magazine* **69**: 47-53.

Strudwick, T. & Lee, P. 2010, *Odynerus simillimus. Hymettus Report* **2010**: 19-29 (unpublished).

Strudwick, T. & Scott, D. 2008. *Odynerus simillimus. Hymettus Report* **2008**: 4-6 (unpublished).

Uffen, R.W.J. 1998. A nest of the wasp *Odynerus spinipes* (L.) (Hymenoptera: Eumenidae) robbed by *Lasius niger* (L.) ants. *British Journal of Entomology and Natural History* **11**: 71.

UK Biodiversity Group. 1999. *Tranche 2 Action plans. Volume IV – invertebrates.* English Nature pp1-473.

Villemant, C., Barbet-Massin, M., Perrard, A., Muller, F., Gargominy, O., Jiguet, F. & Rome. 2011. Predicting the invasion risk by the alien bee-hawking Yellow-legged hornet *Vespa veluntina nigrithorax* across Europe and other continents with niche models. *Biological Conservation* **144**: 2140-2150.

Wright, A. 2007. *Odynerus melanocephalus, O. spinipes. Hymettus Report* **2007**: 15-17 (unpublished).

Wright, A. 2010. *Odynerus melanocephalus. Hymettus Report* **2010**: 13-14 (unpublished).

Yarrow, I. 1954. *Ancistrocerus gazella* (Panzer) (=*A. pictipes* Thomson), an abundant but hitherto undetected Eumenine wasp in Britain. *Journal of the Society for British Entomology* **5**: 78-82.

Index

To superfamilies, genera and species. Main entries and start of sections are given in **bold**. Synonyms are given in *italics*.

Colour plates

Plate 1. Family Mutillidae
Mutilla europaea ♂ 9-14mm

Plate 2. Family Mutillidae
Mutilla europaea ♀ 7-15mm

Plate 3. Family Mutillidae
Smicromyrme rufipes ♂ 4-7mm

Plate 4. Family Mutillidae
Smicromyrme rufipes ♀ 3-6mm

Plate 5. Family Mutillidae
Myrmosa atra ♂ 5-11mm

Plate 6. Family Mutillidae
Myrmosa atra ♀ 4-7mm

Plate 7. Family Sapygidae
Monosapyga clavicornis ♂ 6-10mm

Plate 8. Family Sapygidae
Monosapyga clavicornis ♀ 7-10mm

Plate 9. Family Sapygidae
Sapyga quinquepunctata ♂ 7-11mm

Plate 10. Family Sapygidae
Sapyga quinquepunctata ♀ 9-13mm

Plate 11. Family Tiphiidae
Tiphia femorata ♂ 5-11mm

Plate 12. Family Tiphiidae
Tiphia femorata ♀ 7-14mm

Plate 13. Family Tiphiidae
Tiphia minuta ♀ 5-7mm

Plate 14. Family Tiphiidae
Methocha articulata ♂ 7-11mm

Plate 15. Family Tiphiidae
Methocha articulata ♀ 3-9mm

Plate 16. Family Vespidae
Ancistrocerus antelope ♂ 10-13mm

Plate 17. Family Vespidae
Ancistrocerus antelope ♀ 13-16mm

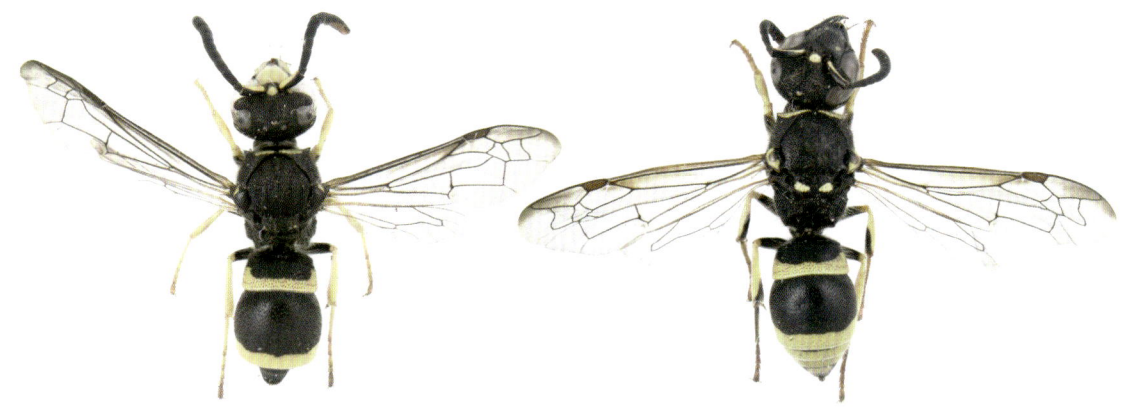

Plate 18. Family Vespidae
Ancistrocerus gazella ♂ 7-9mm

Plate 19. Family Vespidae
Ancistrocerus gazella ♀ 8-12mm

Plate 20. Family Vespidae
Ancistrocerus nigricornis ♂ 6-10mm

Plate 21. Family Vespidae
Ancistrocerus nigricornis ♀ 9-13mm

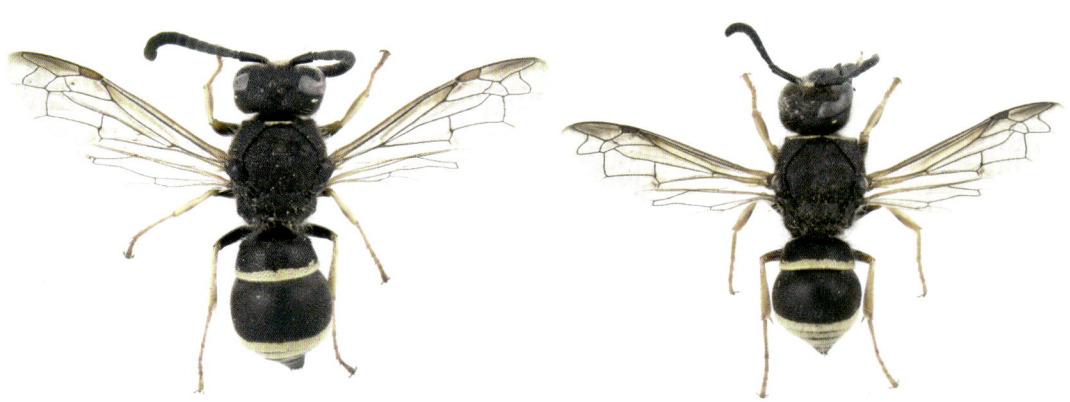

Plate 22. Family Vespidae
Ancistrocerus oviventris ♂ 6-12mm

Plate 23. Family Vespidae
Ancistrocerus oviventris ♀ 11-14mm

Plate 24. Family Vespidae
Ancistrocerus parietinus ♂ 9-12mm

Plate 25. Family Vespidae
Ancistrocerus parietinus ♀ 12-15mm

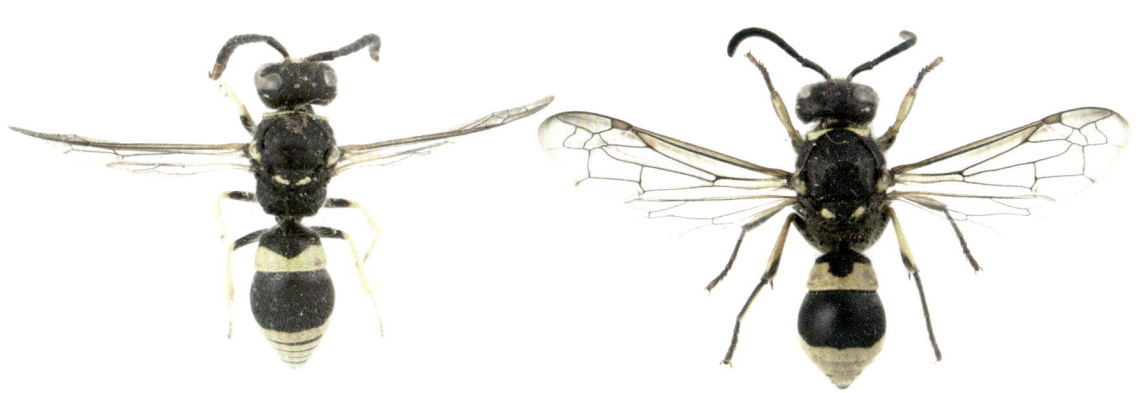

Plate 26. Family Vespidae
Ancistrocerus parietum ♂ 7-11mm

Plate 27. Family Vespidae
Ancistrocerus parietum ♀ 8-12mm

Plate 28. Family Vespidae
Ancistrocerus scoticus ♂ 7-9mm

Plate 29. Family Vespidae
Ancistrocerus scoticus ♀ 10-12mm

Plate 30. Family Vespidae
Ancistrocerus trifasciatus ♂ 7-10mm

Plate 31. Family Vespidae
Ancistrocerus trifasciatus ♀ 10-12mm

Plate 32. Family Vespidae
Eumenes coarctatus ♂ 9-13mm

Plate 33. Family Vespidae
Eumenes coarctatus ♀ 13-15mm

Plate 34. Family Vespidae
Euodynerus quadrifasciatus ♂ 9-11mm

Plate 35. Family Vespidae
Euodynerus quadrifasciatus ♀ 11-12mm

Plate 36. Family Vespidae
Gymnomerus laevipes ♂ 9-10mm

Plate 37. Family Vespidae
Gymnomerus laevipes ♀ 8-11mm

Plate 38. Family Vespidae
Microdynerus exilis ♀ 6-8mm

Plate 39. Family Vespidae
Odynerus melanocephalus ♂ 10mm

Plate 40. Family Vespidae
Odynerus melanocephalus ♀ 8-10mm

Plate 41. Family Vespidae
Odynerus spinipes ♂ 10-12mm

Plate 42. Family Vespidae
Odynerus spinipes ♀ 10-12mm

Plate 43. Family Vespidae
Pseudopiona herrichi ♂ 9-11mm

Plate 44. Family Vespidae
Pseudopiona herrichi ♀ 11-13mm

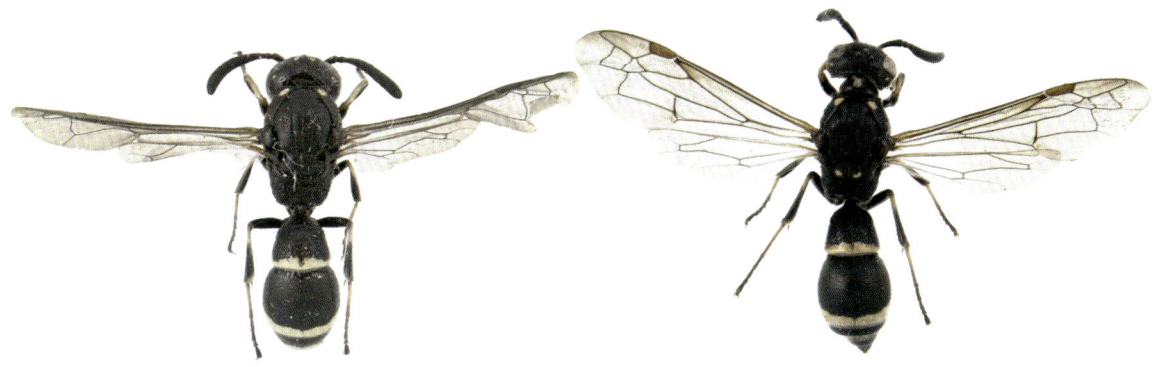

Plate 45. Family Vespidae
Symmorphus bifasciatus ♂ 6-9mm

Plate 46. Family Vespidae
Symmorphus bifasciatus ♀ 7-10mm

Plate 47. Family Vespidae
Symmorphus crassicornis ♂ 8-12mm

Plate 48. Family Vespidae
Symmorphus crassicornis ♀ 10-15mm

Plate 49. Family Vespidae
Symmorphus gracilis ♂ 7-10mm

Plate 50. Family Vespidae
Symmorphus gracilis ♀ 8-12mm

Plate 51. Family Vespidae
Dolichovespula media ♀ 17-21mm

Plate 52. Family Vespidae
Dolichovespula norwegica ♀ 15-17mm

Plate 53. Family Vespidae
Dolichovespula saxonica ♀ 15-18mm

Plate 54. Family Vespidae
Dolichovespula sylvestris ♀ 16-19mm

Plate 55. Family Vespidae
Vespa crabo ♀ 26-30mm

Plate 56. Family Vespidae
Vespula austriaca ♀ 16-18mm

Plate 57. Family Vespidae
Vespula germanica ♀ 16-19mm

Plate 58. Family Vespidae
Vespula rufa ♀ 17-18mm

Plate 59. Family Vespidae
Vespula vulgaris ♀ 15-18mm

Plate 60. *Pseudepipona herrichii*, Hartland Moor. Photo by Robin Williams.

Plate 61. Hornet (*Vespa crabro*) emerging. Photo by Robin Williams.